CRYSTAL OSCILLATOR DESIGN AND TEMPERATURE COMPENSATION

CRYSTAL OSCILLATOR DESIGN AND TEMPERATURE COMPENSATION

Marvin E. Frerking

 VAN NOSTRAND REINHOLD COMPANY
NEW YORK   CINCINNATI   ATLANTA   DALLAS   SAN FRANCISCO
LONDON   TORONTO   MELBOURNE

The author assumes no responsibility for the use of any circuitry, program, or technique contained in this publication nor is any patent license implied.

Van Nostrand Reinhold Company Regional Offices:
New York   Cincinnati   Atlanta   Dallas   San Francisco

Van Nostrand Reinhold Company International Offices:
London   Toronto   Melbourne

Library of Congress Catalog Card Number: 77-17876
ISBN: 0-442-22459-1

Manufactured in the United States of America

Published by Van Nostrand Reinhold Company
135 West 50th Street, New York, N.Y. 10019
Published simultaneously in Canada by Van Nostrand Reinhold Ltd.

15 14 13 12 11 10 9 8 7 6 5 4 3 2

Library of Congress Cataloging in Publication Data

Frerking, Marvin E
   Crystal oscillator design and temperature compensation.

   Includes bibliographical references and index.
   1. Oscillators, Crystal.   2. Oscillators, Transistor.
3. Frequency stability.   4. Temperature control.
I. Title.
TK7862.07F73   621.3815'33   77-17876
ISBN 0-442-22459-1

Foreword

Crystal oscillators have been in use now for well over 50 years—one of the first was built by W. G. Cady in 1921. Today, millions of them are made every year, covering a range of frequencies from a few Kilohertz to several hundred Megahertz and a range of stabilities from a fraction of one percent to a few parts in ten to the thirteenth, with most of them, by far, still in the range of several tens of parts per million.Their major application has long been the stabilization of frequencies in transmitters and receivers, and indeed, the utilization of the frequency spectrum would be in utter chaos, and the communication systems as we know them today unthinkable, without crystal oscillators.

With the need to accommodate ever increasing numbers of users in a limited spectrum space, this traditional application will continue to grow for the foreseeable future, and ever tighter tolerances will have to be met by an ever larger percentage of these devices.

Narrowing the channel spacing—with its concomitant requirements for increasingly more stable carrier frequencies—is but one of the alternatives to increase the number of potential users of the frequency spectrum. Subdividing the time during which a group of users has access to a given channel is another; and many modern radio transmission systems make use of this principle. Here again, the crystal oscillator plays a dominant role; not to control the carrier frequency, but to keep the time slots for the various users coordinated, that is, to serve as the clock rate generator of the systems clocks in transmitters and receivers. The demands on oscillator performance are often even more stringent in this application than for carrier stabilization alone.

The use of crystal oscillators as clock rate generators has seen a rate of growth in the recent past that is nothing short of explosive, with no end in sight yet, in applications that are quite unrelated to the communications field, such as in the quartz wrist watch and in the microprocessor. Other uses include reference standards in frequency counters and time interval meters, gauges for temperature and pressure, and instruments for the measurement of mass changes for scientific and environmental sensing purposes, to name just a few.

In short, the crystal oscillator is now more in demand than ever, and the need for improved performance in mass producible devices becomes more urgent with nearly every new application. An increasing number of engineers, therefore, find themselves confronted with the challenge of designing crystal oscillators with near optimum performance, as tailored to a specific application. Those new in the field are bound to discover very soon that there is no substitute for a considerable amount of hands-on experience. Rarely can a circuit reported on in the literature be used without modifications, and details, not discussed fully in the descriptions provided, are often found to be significant.

The possible combinations of circuit elements that make a viable oscillator are nearly limitless, and while most experienced designers have gravitated toward a few basic configurations, no one circuit, or even small group of circuits, has as yet evolved that is, in all details, universally suitable. Nor does it appear likely that this will happen in the foreseeable future, if for no other reason than because new active devices are continually being brought to market, with often significant advantages for use in oscillator circuits, but requiring different conditions for proper operation. The general principles of crystal oscillator design, however, remain.

The basic building blocks of a crystal oscillator are the feedback circuit containing the crystal unit; the amplifier containing one or more active devices; and circuitry or devices such as needed for modulation, temperature compensation or control, etc.

What is needed most by the circuit designer is a clear approach to understanding the interrelationship of the various circuit elements within each of these building blocks and of the blocks with one another. And this holds true whether the goal is an oscillator, hand-tailored in small quantities to achieve the highest performance possible, or mass produced and capable of meeting the specified requirements under worst case conditions. While such approaches do exist, their exposition in the literature is scarce. It is to fill this void that Mr. Frerking has written this book.

M. E. Frerking is surely one of the most accomplished and innovative practitioners of the art of crystal oscillator design, with extensive experience in the development of high performance oscillators for high volume use. In his book he shares with the reader the design techniques that he has found most useful and conveys a wealth of practical information that will be of immediate use to engineers who are faced with the challenge of designing a crystal oscillator for today's more demanding applications.

Mr. Frerking's book is a timely, and most welcome, major addition to the literature on crystal oscillators.

Erich Hafner, PʜD.
Supervisory Research Physicist
US Army Electronics Technology & Devices Laboratory
Fort Monmouth, NJ

Acknowledgments

The author wishes to express his sincere appreciation to the Collins Telecommunications Products Division of Rockwell International whose support and facilities made this work possible, and to those engineers working in the field of frequency control whose contributions have resulted in many of the techniques described in this book.

The author would also like to acknowledge the assistance rendered by Dr. John Robinson and Mr. H. Paul Brower for their suggestions and assistance in preparing the manuscript.

List of Symbols

Symbol	Description
b_f	forward transfer susceptance
b_i	input susceptance
b_o	output susceptance
b_r	reverse transfer susceptance
C_0	shunt capacity across crystal
C_1	motional arm capacitance
f_a	antiresonant frequency
f_L	frequency at load capacitance C_L
f_s	series resonant frequency
g_f	forward transfer conductance
$g_f(\text{min})$	minimum forward transfer conductance required for oscillation
g_i	input conductance
g_m	forward transfer conductance (transconductance)
g_o	output conductance
g_r	reverse transfer conductance
h_f	forward current transfer ratio
h_i	input impedance
K	Boltzman's constant $1.38 \times 10^{-23}\,\text{J}/^\circ\text{K}$
L_1	motional arm inductance
P_c	power dissipated in crystal
ppm	parts per million
q	electron charge $1.602 \times 10^{-19}\,\text{C}$
R_1	motional arm resistance
R_e	equivalent resistance of crystal
R_{in}	parallel input resistance
R_L	external load resistance
R_{max}	maximum resistance crystal oscillator is capable of handling

R_T total resistive component of collector load

ω_T angular frequency at which the common emitter current
                         gain has decreased to unity

X_e equivalent reactance of crystal

y_f forward transfer admittance

y_i input admittance

y_o output admittance

y_r reverse transfer admittance

Z_f forward transfer impedance

Z_i input impedance

Z_o output impedance

Z_r reverse transfer impedance

Contents

CRYSTAL
OSCILLATOR
DESIGN
AND
TEMPERATURE
COMPENSATION

1
Introduction

The increasing demand in radio communications for channel space as well as the use of sophisticated navigation systems and data transmission has resulted in increased frequency stability requirements in many items of equipment. As a result, the demands on crystal oscillators have become more stringent. In many cases, it is no longer sufficient merely to use a crystal oscillator; now it is necessary to take measures to ensure that the crystal oscillator will possess a high degree of frequency stability. Designs of this type are often quite difficult for the engineer who has had little or no prior experience with crystal oscillators; consequently, much of the material in this book is directed to the individual who has a good background in circuit theory but who is not necessarily experienced with crystal oscillator design.

The book deals primarily with transistor oscillators, since nearly all precision oscillators at the present time use discrete transistors. The use of gate oscillators and clock oscillator integrated circuits is widespread in lower stability applications, and these are discussed in Chapter 7.

A practical treatment of quartz crystal resonators is presented in Chapter 5 which gives the designer a good working knowledge of the devices. In Chapter 6, the nonlinear properties of transistors are explored to enable prediction of the amplitude of oscillation and the harmonic content for various oscillators. Chapter 7 then brings together all the information already presented and presents the actual design equations for oscillators covering the entire frequency spectrum from several kHz to 150 MHz. It also includes over 20 tested circuits with component values.

Crystal oscillators, in general, are more critical than most electronic circuits. As such, it behooves the design engineer to take special precautions to ensure that his oscillator circuit will perform properly when produced in quantity. Chapter 8 consists of a discussion of several tests which should be conducted to determine with reasonable assurance whether the circuit will perform properly when produced in quantity.

Crystal ovens, discussed briefly in Chapter 9, are used almost exclusively to achieve stabilities better than 5×10^8. The treatment of ovens is limited primarily to a description of the basic techniques and what can be achieved, no attempt is made to give detailed design information.

The spectral purity of crystal oscillators may be an important consideration in some applications and, although not treated in this book, should not be overlooked. The reader is directed to numerous articles in the literature for designs of this type.

A unique system is presented in Chapter 10, whereby a microprocessor can be used to temperature-compensate a crystal oscillator. This system is compared with three other methods for temperature compensation. Chapter 10 also contains a thorough treatment of temperature compensation in general, which enables the average design engineer to accomplish successful compensation of semiprecision oscillators, improving the stability by as much as two orders of magnitude.

Many of the derivations required to develop the design equations are carried out in the appendices, but the conclusions are presented in the main text. This results in an easily readable volume with the details still available for those interested in probing deeper into the mechanics of the derivations and the assumptions made.

2
Basic Oscillator Theory

In undertaking the design of a crystal oscillator, an understanding of basic oscillator principles is not only desirable but essential. Therefore, a brief explanation of the operation of a crystal oscillator is given here. Basically, a crystal oscillator can be thought of as a closed loop system composed of an amplifier and a feedback network containing the crystal. Amplitude of oscillation builds up to the point where nonlinearities decrease the loop gain to unity. The frequency adjusts itself so that the total phase shift around the loop is 0 or 360 degrees. The crystal, which has a large reactance–frequency slope, is located in the feedback network at a point where it has the maximum influence on the frequency of oscillation. A crystal oscillator is unique in that the impedance of the crystal changes so rapidly with frequency that all other circuit components can be considered to be of constant reactance, this reactance being calculated at the nominal frequency of the crystal. The frequency of oscillation will adjust itself so that the crystal presents a reactance to the circuit which will satisfy the phase requirement. If the circuit is such that a loop gain greater than unity does not exist at a frequency where the phase requirement can be met, oscillation will not occur.

The application of these principles to oscillator design usually is difficult because many factors play an important part in the operation. As a result, the design of transistorized crystal oscillators is often a "cut and try" procedure.

Methods have been developed for predicting the amplitude of oscillation based on the small-signal loop gain. The reduction in gain for a transistor operating at large signal values is predictable and has been plotted as a function of the ac base-to-emitter voltage. Since it is known that the loop gain after equilibrium has been reached will be unity, the reduction factor is numerically equal to the small-signal loop gain. Using this value, the amplitude of oscillation can be predicted from the graphs.

3

3
Methods of Design

Three methods of design are presented in this book, each of which has its advantages. The first, which is highly experimental, consists of giving a qualitative explanation of how the circuit works and presenting a number of typical schematic diagrams for that oscillator configuration. The second method consists of deriving the equations for oscillation in terms of the Y-parameters of the transistor. The third method consists of measuring the gain and input impedance of the transistor as a function of its load impedance. This information is used to calculate component values for the circuit with relatively simple equations. The amplitude of oscillation can then be predicted using the methods of paragraph 3.4.

3.1. EXPERIMENTAL METHOD OF DESIGN

The experimental method of design consists of finding a suitable circuit which can be modified and/or optimized to meet a particular set of requirements. To assist in this design approach, Chapter 7 contains a number of laboratory tested oscillator circuits and a qualitative explanation of their operation. The appropriate circuit type most suited for a particular application can be selected with the aid of Table 7-1. The individual circuits have not been designed or optimized with respect to any particular performance characteristic, but sufficient reserve gain has been provided to allow some modification.

The following precautionary items must be presented in regard to the use or modification of any of these circuits:

a. Since the mechanical arrangement of a circuit usually affects its performance, complete testing of the circuit in accordance with Chapter 8 should be accomplished even though the circuit values presented are used.

b. Substitution of transistors or gates for those specified should be within the same basic family and power level. Indiscriminate substitution of active element types may greatly change the performance of a given oscillator circuit.

3.2. Y-PARAMETER METHOD OF DESIGN

The second approach to oscillator design consists of using the Y-parameters of the transistor (see Chapter 6). The equations for oscillation are derived in the following manner. Using the block diagram of Figure 3-1, the complex equation for oscillation can be shown (see Appendix A) to be:

$$y_f Z_f + y_i Z_o + y_o Z_i + y_r Z_r + \Delta y \, \Delta Z + 1 = 0, \qquad (3\text{-}1)$$

where

$$\Delta y = y_o y_i - y_f y_r, \qquad \Delta Z = Z_o Z_i - Z_f Z_r.$$

Although any set of parameters may be used for the amplifier and any set for the feedback network, it is convenient to use Y-parameters for the amplifier and Z-parameters for the feedback network. . . . It is important to note that the use of equation (3-1) implies the assumption that the amplifier is a linear circuit. The application of equation (3-1) therefore can yield no information concerning harmonic generation or the limiting of amplitude as the result of dependence of circuit parameters upon amplitude. The assumption that the amplifier is linear is not valid at large amplitudes. At large amplitudes, the Y-parameters therefore must be defined as the ratios of fundamental components of current to fundamental components of voltage![41]

The equations for specific oscillator types are derived by determining the Z-parameters of the feedback network and substituting them

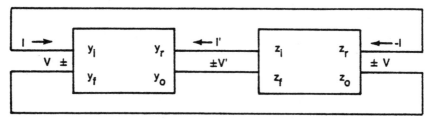

Figure 3-1. Block diagram of a transistorized crystal oscillator.

into equation (3-1). The complex equation is then separated into real and imaginary parts. The real part generally yields an expression for the transconductance g_f required for oscillation while the imaginary part yields an expression for the crystal reactance X_L necessary to satisfy the phase shift requirement. The equations and the assumptions made are presented for the various oscillators in Chapter 7.

Since the equations in general do not give highly accurate results, it is well to use them in connection with the experimental approach (see section 3.1). However, the equations do give an indication of how changing a given component will affect the overall performance and thus are often quite useful. The equations are generally of the form

$$g_f = f_1(a, b, c, d, \ldots)$$
$$X_L = f_2(a, b, c, d, \ldots),$$

where $a, b, c, d \ldots$ represent various components and parameters of the circuit; X_L is the crystal reactance; and g_f is the small-signal transconductance required for oscillation to begin. The ratio g_f (transistor)/g_f (required) is a measure of the loop gain, which must be greater than unity for oscillations to build up. Generally, if the loop gain is greater than 2 to 3, satisfactory operation results. If limiting

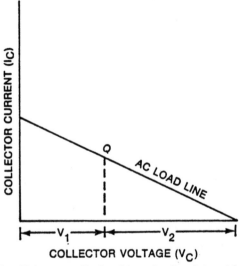

Figure 3-2.   Voltage prediction from the Q-point and load line.[35]

takes place as a result of the base-to-emitter junction being cut off during part of the cycle, the amplitude of oscillation can be predicted using Figure 3-6. Limiting of this type generally results in good frequency stability. The oscillator may be biased to produce collector limiting. If this is the case, the output voltage can be determined by constructing a load line as shown in Figure 3-2. The peak output voltage will be approximately V_1 or V_2, whichever is smaller. This is a rule of thumb only and not highly accurate.

The oscillator should be designed to require the same crystal reactance (X_L) as that called out by the crystal specification for on-frequency operation. In the case of series resonant crystals, $X_L = 0$.

3.3. POWER GAIN METHOD OF DESIGN*

The third approach to oscillator design is basically a power gain analysis. Phase shift considerations are taken care of experimentally by getting the crystal to operate on frequency. The usefulness of this design approach generally is limited to series mode oscillators which can be represented by the block diagram of Figure 3-3.

The power gain required from the transistor must be sufficient to supply the output power, power losses, and the input power required

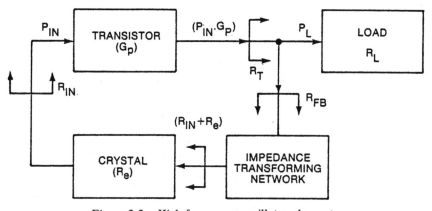

Figure 3-3. High-frequency oscillator elements.

*The results presented here are essentially a summary of the Power Gain Method of design developed under sponsorship of the US Army Electronics Command, see reference 31.

for the transistor:

$$(P_{in} G_P) = P_L + P_{in} + P_d, \qquad (3-2)$$

where

P_{in} = input power to transistor,
G_P = power gain of transistor,
P_L = output power to an external load, and
P_d = all other power losses within the oscillator circuit.

Using equation (3-2), an oscillator may be designed as follows:

A. *Determine the transistor power gain (experimentally).*

Step 1. Connect the transistor as a single-tuned amplifier in the grounded-base or grounded-emitter configuration, whichever is to be used in the type of oscillator being designed. A circuit similar to that of Figure 3-4 may be used. The circuit should be arranged so that it can be mounted on the impedance measuring device such as a network analyzer or *RX* meter with the input near the ungrounded terminal. Provisions should be made for connecting RF voltmeters to the input and output of the transistor.

Step 2. Measure the power gain and input impedance as a function of the load resistance R_T.*

(a) For various values of load resistance, determine the power gain and the input impedance, increasing the value of the load resistor at each step until instability occurs.
(b) Plot the power gain and input resistance versus load resistance. A graph similar to that of Figure 3-5 should result.
(c) From the power gain graph, select a value R_T giving a gain of 200–300, and note the input resistance R_{in} at the power gain selected.

B. *Calculate the feedback network.* Power gain values determined in A include all circuit losses that will be present in the oscillator

---

*The maximum input voltage that can be applied to the transistor before nonlinearity occurs is about 10 mV. Since the output of the Boonton *RX* meter is about 100 mV, it must be modified. The addition of an appropriate level control is described fully in the *RX* meter instruction manual. In this discussion, R_T refers to the total load resistance seen by the collector. R_L, the external load, is included in R_T.

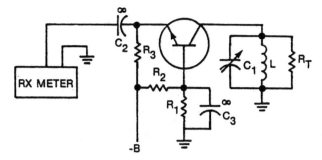

Figure 3-4.   Single-tuned amplifier connection.

except the crystal loss. The crystal loss is included in the following manner.

Step 1. The ratio of the total feedback power to the transistor input power is given by

$$\frac{P_{\text{FB}}}{P_{\text{in}}} \doteq \frac{(R_{\text{in}} + R_e)}{R_{\text{in}}}, \tag{3-3}$$

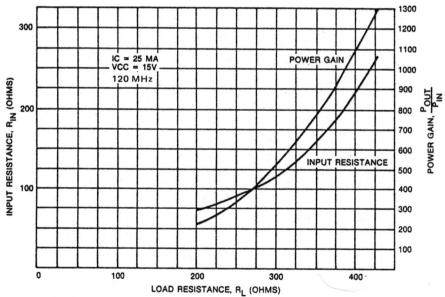

Figure 3-5. Input   resistance   and   power   gain   versus   load   for   2N2218 transistor.[28, 29]

where

P_{FB} = total feedback power,
P_{in} = the power input to the active device,
R_{in} = the input resistance of the transistor, and
R_e = the series resonant resistance of the crystal.

Incorporating this loss due to the crystal into the input circuit gives the modified power gain G_p' as

$$G_p' = \frac{G_p R_{\text{in}}}{(R_{\text{in}} + R_e)}. \tag{3-4}$$

Step 2. The next step is to determine the ratio of the output power to the feedback power. All losses are accounted for now; therefore,

$$P_o = P_L + P_{\text{FB}}. \tag{3-5}$$

The output power also is given by

$$P_o = (P_{\text{FB}} \, G_p'). \tag{3-6}$$

Combining equations (3-5) and (3-6) gives

$$(P_{\text{FB}} \, G_p') = P_L + P_{\text{FB}}, \tag{3-7}$$

or

$$P_{\text{FB}} = \frac{P_L}{(G_p' - 1)}. \tag{3-8}$$

P_{FB} can be represented by an equivalent resistor R_{FB} (whose power dissipation is P_{FB} placed in parallel with the external load R_L. R_L and R_{FB} are subjected to the same voltage; therefore, the resistance ratio is inverse to the power ratio, and

$$R_{\text{FB}} = R_L (G_p' - 1). \tag{3-9}$$

Now R_T, the total load resistance, is the parallel combination of R_{FB} and R_L; using this with equation (3-9) and rearranging terms gives

$$R_L = \frac{R_T \, G_p'}{(G_p' - 1)} \tag{3-10}$$

and

$$R_{FB} = R_T G'_p. \tag{3-11}$$

Using equations (3-10) and (3-11), the values of R_{FB} and R_L can be determined. The use of a G'_p of one-third to one-half the value determined from equation (3-4) should provide an adequate feedback power safety factor.

Step 3. The last step in the procedure is the determination of the required impedance transformation ratio of the feedback circuit. This is the ratio of R_{FB} to $(R_{in} + R_e)$ or

$$\text{Required impedance transformation ratio} = \frac{R_{FB}}{(R_{in} + R_e)}. \tag{3-12}$$

There are several types of impedance transforming networks which can be used, e.g., a capacitive tap on the output tuned circuit, a pi network, or a transformer. The properties of specific networks are treated briefly with the discussion of particular oscillator circuits in Chapter 7. Detailed discussions of several feedback networks are given in references 31, 32, and 35.

The power gain approach to the design of crystal oscillators is one of the few approaches simple enough to be of practical value. Accuracy is only fair and the difference from actual oscillator loop gain normally will not exceed 2 or 3. Also, a considerable amount of component value adjusting usually is necessary to get the crystal to operate on frequency. The approach is of the most value in designing oscillators of high frequency and high output power.

In general, the Y-parameter approach is a better design method for low-power oscillators. (If, however, the Y-parameters of a transistor are not known, or if from other considerations the reader elects to use the power gain method, it is suggested that reference 31 be consulted, since only the principles of this approach have been outlined here, and a detailed explanation of each step is given in the reference.)

3.4. NONLINEAR MODIFICATIONS

The small-signal analysis discussed in section 3.2 is valid until the ac base-to-emitter voltage builds up to about 10 mV. For values greater than this, significant changes occur in the forward transconductance as well as the input and output impedances. The magnitude of these changes is derived in Appendix G for the basic transistor and in Appendix H for a transistor with emitter degeneration. If the initial loop gain determined by g_f (transistor)/g_f (required) is calculated, the result can then be used to predict the base-to-emitter voltage, the input and output impedances, the harmonic current, and the bias shift. The curves of Figure 3-6 illustrate the method.

Suppose that the initial loop gain is 3. After the amplitude of oscillation has built up to its equilibrium value, the actual loop gain will be unity. Therefore the ratio g_m/g_{mo} must be 0.333. From Figure 3-6 we see that V, the normalized ac base-to-emitter voltage, will be 5.7. The actual base voltage is then 5.7 KT/q where

K = Boltzman's constant, 1.38×10^{-23} J/$^\circ$K;
q = electron charge, 1.602×10^{-19} C; and
T = temperature in $^\circ$K.

At room temperature $KT/q \doteq 26$ mV; therefore, the actual voltage is 5.7 $\times$ 26 = 148.2 mV.

Once the base voltage is known, it is normally fairly straightforward to calculate the voltage in any other part of the circuit.

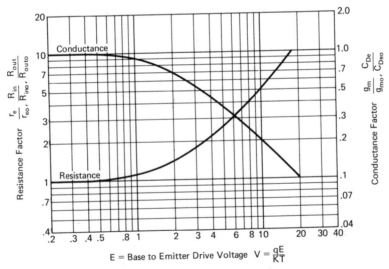

E = Base to Emitter Drive Voltage $V = \dfrac{qE}{KT}$

Figure 3-6.   Transistor parameters versus signal voltage.

It should be noted that the input capacitance is reduced by the same ratio as the small-signal loop gain, which results in a slight increase in frequency.

Curves showing the harmonic currents as well as the bias shift are presented in Chapter 6. They may be useful in predicting the performance of the oscillator if a harmonic of the fundamental frequency is used.

Specific nonlinear equations, based on the principle of harmonic balance, are also derived for the Colpitts oscillator in Appendix I, and the results are presented in section 7.3.

4

Oscillator Frequency Stability

The term *frequency stability* is a generic term which means a variety of things to different people depending on their individual interests. In its broadest concept, it means the degree of constancy of the frequency of an oscillator under a particular set of conditions. In crystal oscillator applications, there are several different types of frequency stability:

  a. Frequency stability as affected by environmental changes consisting primarily of temperature, voltage, and load variations.
  b. Long-term frequency drift as affected by aging of the quartz crystal resonator.
  c. Short-term frequency stability or phase stability.

Frequency stability is used in this book to mean either a or b, both of which will be discussed in this chapter. In some specifications frequency stability is meant to be the sum of a and b; however, because so much misunderstanding has resulted, it is recommended that when this usage is adopted it should be clearly stated that frequency stability is meant to include both environmental stability and aging for a specified time period. The term frequency accuracy is also sometimes specified and is a measure of the actual frequency compared to an established standard. It results from the initial setting error and the stability.

4.1. TEMPERATURE EFFECTS ON FREQUENCY

The frequency of a crystal oscillator is affected by changes in ambient temperature. These changes in temperature can affect the value of any of the components which comprise the oscillator circuit. If these component variations do not cancel each other, a change in the nominal operating frequency of the oscillator will result. The frequency determining component most severely affected by any temperature change is the quartz crystal. This effect is shown graphically for AT-cut crystals in Chapter 5, Figure 5-6. (A discussion of the temperature coefficients of other crystal cuts can be found in reference 6.)

In some applications, sufficient frequency stability can be obtained from the quartz crystal. The limit obtainable over the full military temperature range of $-55°C$ to $+105°C$ with an AT-cut crystal is approximately $±0.002$ percent. This limit can be improved within a reduced temperature range. Many applications require stabilities considerably better. In such cases, two methods are available for eliminating or reducing the effects of temperature changes on the crystal oscillator, namely, temperature control and temperature compensation.

4.1.1. Temperature Control

The degree of temperature control required on a particular oscillator is determined primarily by the specifications of the system in which it is to be used. Stabilities of approximately $±5$ parts in 10^7 can be obtained using plug-in crystal ovens with the oscillator circuitry external to the oven. Stabilities to several parts in 10^9 can be obtained with proportionally controlled ovens containing the crystal and oscillator circuitry. (A proportionally controlled oven uses a temperature-controlling system in which the power supplied to the oven is proportional to the heat loss. (Refer to section 9.3.) For stabilities better than 5 parts in 10^9, it is generally necessary to use a two-stage oven. This may be a combination of two ovens with a single control circuit or two independent proportionally controlled ovens.

Crystal ovens have several disadvantages which tend to limit their usage in some applications. These are as follows:

1. A warm-up time is required.
2. The volume is relatively large.
3. The power consumption is high.
4. The reliability of the components in the oven is reduced if the application requires frequent turning on and off of the oven.

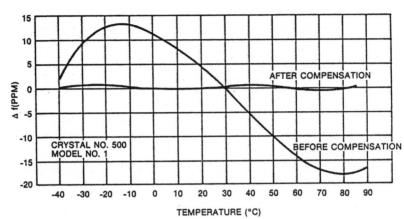

Figure 4-1. Frequency versus temperature characteristic for a typical temperature-compensated crystal oscillator.

4.1.2. Temperature Compensation

Temperature compensation of crystal oscillators is very practical to achieve frequency stabilities in the range of ±10 to ±0.5 ppm. With considerable care, compensation to ±0.05 ppm is possible.

Temperature compensation is generally achieved by placing a voltage variable capacitor in series with the crystal. A voltage is then applied to the capacitor, which pulls the crystal frequency by precisely the amount that it drifted in temperature but in the opposite direction. The voltage is generated either by a thermistor–resistor analog network or by a digital system followed by a digital-to-analog converter.

The means for temperature compensation are discussed in considerable detail in Chapter 10. Figure 4-1 shows the improvement in frequency that was achieved using a three-thermistor analog network in a 3.2-MHz crystal oscillator.

4.2. LONG-TERM FREQUENCY DRIFT

The phrase *long-term frequency drift* usually refers to the gradual drift in average frequency of an oscillator due to aging of components, notably the quartz crystal. It is not meant to include the short-term variations discussed in section 4.3 or the deviations due to ambient temperature change discussed in section 4.1. The aging of a quartz crystal itself is discussed in section 5.7.

4.3. SHORT-TERM FREQUENCY STABILITY

The phrase *short-term frequency stability* refers to changes in the oscillator frequency which result from interaction of the desired signal with an unwanted signal or noise. It is not meant to include frequency variations due to component aging or ambient temperature change. The type of interaction may be simple superposition, amplitude modulation, frequency modulation, phase modulation, or any combination thereof. Only in the case of FM or PM is there a true change in frequency. The other types may cause an apparent change in frequency which may vary with different frequency-measuring techniques. For this reason, the signal-to-noise ratio or sideband level of an oscillator is sometimes specified. If phase modulation is the only type of interaction being considered, or is predominant, the term *phase stability* may be used in place of short-term frequency stability. Frequency modulation and phase modulation are related by the modulation frequency. If the undesirable signal is sinusoidal, this relationship is given by

$$\Delta \theta = \frac{\Delta f}{f_m}$$

where $\Delta \theta$ is the peak phase deviation in radians, Δf is the peak carrier frequency deviation, and f_m is the frequency of the undesirable signal.

As is the case of any FM or PM signal, theoretically an infinite number of sidebands exist. The total phase deviation is usually so small with crystal oscillators, however, that only the first pair of sidebands is significant. The relationship between these sidebands and the phase deviation is given in Figure 4-2. This graph does not consider the presence of AM. The mathematical development of Figure 4-2 is given in Appendix J. In the case of noise modulation, the sideband levels are often specified in decibels below the carrier per hertz of bandwidth (dB/Hz). For a narrow-bandwidth measurement system, pure FM or PM noise modulation results in the same sideband level as shown in Figure 4-2. Here the sideband level is in-

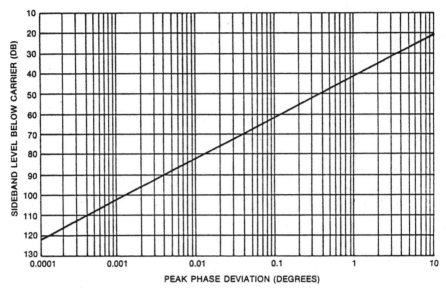

Figure 4-2. Sideband level versus phase deviation.

terpreted to be the ratio of the rms value of the noise sideband to the rms value of the carrier, and the abscissa is $\sqrt{2}$ times the rms phase deviation. For noise simply added to the signal, the sidebands are uncorrelated and the apparent phase deviation is 3 dB lower for the same sideband level. The rms phase jitter is then given by $\Delta\theta = 10^{-dB/20}$ rad rms, where dB refers to the level of either the upper or lower sideband in a bandwidth numerically equal to the baseband in which the phase jitter is measured.*

In many cases short-term frequency stability is best specified in the time domain and is given as the rms fractional frequency deviation for some specified measurement time τ. For example a precision crystal oscillator might exhibit a short-term stability of 1×10^{-11} rms for one-second averaging times. If a large number of frequency measurements, say n, are made using an averaging time of τ seconds, the standard deviation can be computed using the statistical relationship

---

*Short-term frequency stability and/or phase noise can be conveniently measured using a phase-locked loop with two identical oscillators, with a spectrum analyzer, a computing frequency counter, or a frequency stability analyzer.

$$\sigma_n^2(\Delta f) = \frac{1}{n-1} \left[\sum_{i=1}^{n} (f_i)^2 - \frac{1}{n} \left(\sum_{i=1}^{n} f_i \right)^2 \right].$$ (4-1)

It has been found, however, that for large numbers of measurements, the elapse time is so large that frequency aging and temperature effects tend to influence the results and σ becomes a function of how long the test was run. A better method and one which has become standard is to use the Allan variance. In using this method individual variances are computed from adjacent pairs of frequency readings and the average of the variances forms the basis for the definition. Taking $n = 2$, equation (4-1) simplifies to

$$\sigma_2^2(\Delta f) = \frac{(f_1 - f_2)^2}{2}.$$ (4-2)

The frequency stability is then found by taking the square root of the average of the variances, and is

$$\sigma_y(\tau) = \left[\frac{1}{2N} \sum_{i=1}^{N} (f_{2i} - f_{2i-1})^2 \right]^{1/2}$$ (4-3)

where τ is the measurement time for each frequency reading with no dead time between readings, and N is the number of measurement pairs used. A fairly large number of readings is required to compute a reliable value of $\sigma_y(\tau)$, and $N = 100$ is quite common.

The time domain method of specifying short-term frequency stability is useful for counting intervals ranging from less than a millisecond to about 100 seconds. It is possible to convert from time domain measurements to frequency domain performance and vice versa. Indeed this is a very powerful method of determining the frequency spectral content of an oscillator within a fraction of 1 Hz of the carrier.[51] In general, however, it is best to specify the characteristic which is actually important to the system. For example if it is the phase stability that is important than this should be specified.

The art of designing oscillators for best short-term frequency stability is not treated in this book; however, it should be pointed out that it is a very important consideration in the design of some oscillators. A rigorous definition of short-term frequency stability is itself quite complex, and the reader is referred to reference 52 for a comprehensive treatment of the subject.

5
Quartz Crystal Resonators

The importance of quartz crystal resonators in electronics results from their extremely high Q, relatively small size, and excellent temperature stability.

A quartz crystal resonator utilizes the piezoelectric properties of quartz. If a stress is applied to a crystal in a certain direction, electric charges appear in a perpendicular direction. Conversely, if an electric field is applied, it will cause mechanical deflection of the crystal. In a quartz crystal resonator, a thin slab of quartz is placed between two electrodes. An alternating voltage applied to these electrodes causes the quartz to vibrate. If the frequency of this voltage is very near the mechanical resonance of the quartz slab, the amplitude of the vibration will become very large. The strain of these vibrations causes the quartz to produce a sinusoidal electric field which controls the effective impedance between the two electrodes. This impedance is strongly dependent on the excitation frequency and possesses an extremely high Q.

Electrically, a quartz crystal can be represented by the equivalent circuit of Figures 5-1 and 5-2 where the series combination R_1, L_1, and C_1 represent the quartz, and C_0 represents the shunt capacitance of the electrodes in parallel with the holder capacitance. The inductor L_1 is a function of the mass of the quartz, while C_1 is associated with its stiffness. The resistor R_1 results from the loss in the quartz and in the mounting arrangement. The parameters of the equivalent circuit can be measured quite accurately using the crystal impedance (CI) meters,* vector voltmeters,[14,49] or bridge measurement tech-

---

*RFL Industries, Boonton, NJ, Model 5950, with plug-in units to cover frequency of interest. Old crystal impedance meters are TS-710/TSM, 10–1100 kHz, TS-630/TSM, 1–15 MHz; TS-683/TSM, 10–140 MHz; and AN/TSM-15, 75–200 MHz.

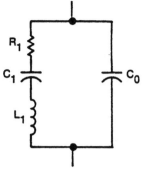

Figure 5-1. Simplified diagram of the equivalent circuit of a quartz crystal

niques.[19]   A reactance-frequency plot of the equivalent circuit is given in Figure 5-3, and a reactance–resistance plot is given in Figure 5-4. The portions circled on these figures are expanded in Figure 5-5.

Several equations have been derived in Appendix K which are useful when using the equivalent circuit. The results are presented below. Several frequencies are marked in Figures 5-4 and 5-5. The first of these is f_s. This is the frequency at which the crystal is series resonant, and is given by

$$f_s = \frac{1}{2\pi\sqrt{L_1 C_1}} \qquad (5\text{-}1)$$

where

f_s = series resonant frequency in hertz,
L_1 = motional arm inductance in henrys, and
C_1 = motional arm capacitance in farads.

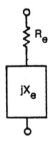

Figure 5-2. Impedance representation of a quartz crystal.

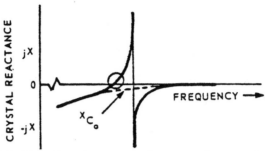

Figure 5-3. Plot of reactance versus frequency for a quartz crystal.

The second point, f_r, represents the frequency at which the crystal appears purely resistive ($X_e = 0$). Point f_r is different from f_s only because of the presence of C_0, and for practical purposes can be considered equal to f_s. The third point labeled, f_L, is the frequency at which the crystal is antiresonant with a given external capacitor C_L. If Δf is the frequency shift ($f_L - f_s$) between series resonance and this load point, then

$$\frac{\Delta f}{f_s} = \frac{C_1}{2(C_0 + C_L)} \tag{5-2}$$

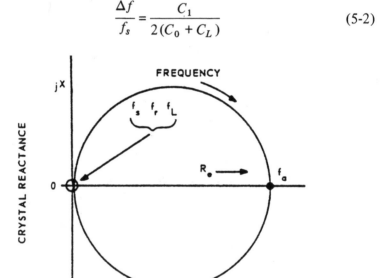

Figure 5-4. Plot of reactance versus resistance for a quartz crystal.

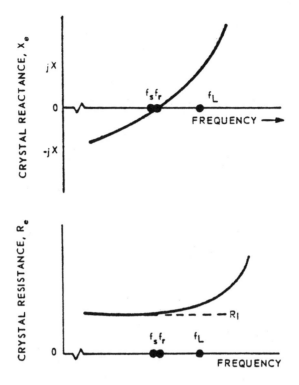

Figure 5-5. Expanded portions of crystal reactance (A) and resistance, (B). (See Figures 5-3 and 5-4).

where

Δf = frequency shift

$(f_L - f_s)$ in hertz,

C_1 = motional arm capacitance in picofarads,

C_0 = crystal holder capacitance in picofarads, and

C_L = external load capacitance in picofarads.

The point labeled f_a is the antiresonant frequency of the crystal with its own holder capacitance C_0. It is given by

$$f_a = f_s \left[1 + \frac{C_1}{2C_0} \right]$$

(5-3)

where

f_a = antiresonant frequency in hertz,
f_s = series resonant frequency in hertz,
C_1 = motional arm capacitance in picofarads, and
C_0 = crystal holder capacitance in picofarads.

Furthermore, it can be shown (see Appendix K) that the equivalent resistance R_e in the region between series (f_s) and antiresonance (f_a) is given by

$$R_e = R_1 \left(\frac{C_L + C_0}{C_L} \right)^2 \tag{5-4}$$

provided the assumption $\left| X_{C_0} \left(\dfrac{C_L}{C_0 + C_L} \right) \right| \gg R_1$ is true, where

$$X_{C_0} = -\frac{1}{2\pi f C_0}.$$

Normally a crystal is operated between its series resonant frequency and its antiresonant frequency so that the reactance X_e is either zero or inductive.

To help the engineer acquire a practical grasp of the equivalent circuit, Table 5-1 is included to give a rough idea of the magnitude of the various equivalent circuit components.

The parameters of a quartz crystal resonator may be varied greatly by the angle at which the crystal blank is cut from the raw quartz and by the mode of vibration. This is primarily a concern of the crystal manufacturer and will not be discussed in detail here. (An excellent

Table 5-1. Typical Crystal Parameter Values.

Parameters	200-kHz[31] fundamental	2-MHz[31] fundamental	30-MHz[31] third overtone	90-MHz fifth overtone
R_1	2 kΩ	100 Ω	20 Ω	40 Ω
L_1	27 H	520 mH	11 mH	6 mH
C_1	0.024 pF	0.012 pF	0.0026 pF	0.0005 pF
C_0	9 pF	4 pF	6 pF	4 pF
Q	18×10^3	54×10^3	10^5	85×10^3

treatment of crystal cuts is given in reference 6.) Several properties of crystal resonators are of concern to the designer of crystal oscillators and will be discussed in the following paragraphs.

5.1 LOAD CAPACITANCE

From Figures 5-3, 5-4, and 5-5 it can be seen that the frequency of the crystal will vary to some extent depending upon the reactance that the crystal must present to an external circuit. Since the frequency difference between series and antiresonance ($f_a - f_s$) may be on the order of 1 percent for some crystals, it is important that the crystal be ground to frequency at the load reactance value with which it will be used in the oscillator. Four load conditions have become standard and are nearly always used. With the first two of these, the crystal acts like an inductive reactance which will resonate with either 30 or 32 pF at the operating frequency. Hence, the load capacitance $C_L = 30\,pF$ or $C_L = 32$ pF. Crystals of this type must be used in parallel resonant oscillators. A second common load point is series resonance, where the crystal acts like the resistor R_1. Crystals of this type must be used with series resonant oscillators. A fourth load point, $C_L = 20$ pF, is sometimes used for crystals below 500 kHz.

5.2 PIN-TO-PIN CAPACITANCE

Pin-to-pin capacitance (C_0 of Figure 5-1) refers to the capacity of the electrodes on the quartz as well as that of the holder itself. The holder capacitance is usually around 0.5 pF and the remaining capacitance is due to the electrodes plated on the quartz. C_0 should be restricted to about 5 pF for AT-cut* crystals while it may be somewhat higher for low-frequency cuts. It becomes important to minimize C_0 for VHF crystals, where it may cause the oscillator to free-run (to oscillate not crystal-controlled). C_0 may be reduced in crystal manufacture by reducing the electrode spot size on the crystal blank. However, this tends to increase the resistance R_1.

---

*The AT-cut is the basic high-frequency crystal normally used in the range from 1 to 150 MHz.

5.3 RESISTANCE

The resistance of a crystal is specified at the rated load capacitance, although this usually does not differ grossly from the series resistance R_1. The maximum allowable resistance for a given crystal type may vary from about 40 Ω for VHF crystals to approximately 500 kΩ for audio-frequency crystals. It is important to make certain that an oscillator will function properly with a crystal of the maximum specified resistance.

5.4 RATED OR TEST DRIVE LEVEL

Drive level refers to the power dissipated in the crystal. Rated or test drive level is the power at which all requirements of the crystal specification must be met. The drive level specification should reasonably duplicate the actual drive level at which the crystal will be used because frequency is somewhat dependent on drive level. AT-cut crystals generally can withstand a considerable overdrive without physical damage; however, the electrical parameters are degraded at excessive drive. Low-frequency crystals (especially flectural mode crystals) may fracture if overdriven. Drive level ratings vary from 5 μW below 100 kHz to about 10 mW in the 1- to 20-MHz region for fundamental mode crystals. Overtone crystals which are generally used above 20 or 30 MHz are often rated at 1–2 mW of drive.

5.5 FREQUENCY STABILITY

The frequency stability of a crystal generally is limited by its temperature coefficient and aging rate. AT-cut crystals have a better temperature coefficient than most other cuts. Common frequency tolerance specifications are ±0.005 percent or ±0.0025 percent from –55°C to +105°C. These include calibration tolerance; thus, the actual temperature coefficient is slightly better. Improved temperature coefficients can be obtained if the temperature range is limited. This can be seen in Figure 5-6, which gives frequency–temperature curves for AT-cut crystals. These curves may be represented by cubic equations and are strongly dependent on the angle of cut of the quartz blank from the mother crystal. The points of zero temperature coefficient

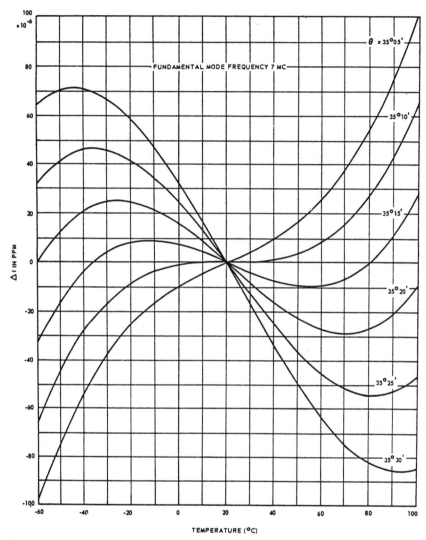

Figure 5-6. Frequency–temperature–angle characteristics of plated AT-type natural quartz crystal resonators.[4]

are called the turning points (lower and upper turning-point temperatures). One turning point can be placed where desired by selecting the angle of cut; the other turning point then is determined, since the turning points are symmetrical about a point in the 20–30°C range.

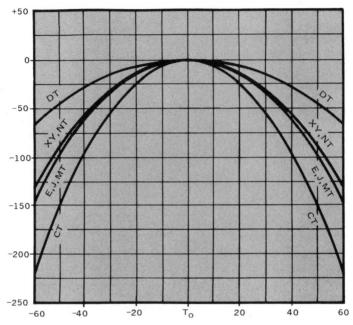

Figure 5-7. Frequency–temperature characteristics of low-frequency crystal cuts. (Courtesy Northern Engineering Laboratories)

The slope between the turning points becomes smaller as the turning points move together. Crystals designed for use in an oven should be cut so that a turning point occurs at the oven temperature. Figure 5-7 shows the frequency–temperature curves for several low-frequency cuts. The J-cut is used below 10 kHz, while an XY-cut may be used from about 3 kHz to 85 kHz. An NT-cut may be used in the 10 kHz to 100 kHz range. A DT-cut is applicable from 100 kHz to about 800 kHz and a CT from perhaps 300 kHz to 900 kHz.

5.6. FINISHING OR CALIBRATION TOLERANCE

Finishing tolerance is the maximum allowable error in frequency of a crystal at some specified temperature. If ±0.005-percent AT-cut crystals are used, it is often desirable to specify a room temperature finishing tolerance of, e.g., ±0.0015 percent so that oscillators can be tuned conveniently to frequency in production. If oscillators are to be tuned to frequency, the finishing tolerance must be less than the tuning range of the oscillator. In the case of temperature controlled crystals, the finishing tolerance is specified at the nominal operating temperature of the oven. Another use of finishing tolerance is with an alternative method of specifying the overall frequency tolerance of a crystal. It is sometimes desirable to specify a room temperature finishing tolerance and a maximum deviation from the room temperature frequency over the temperature range. This method may be used in place of specifying a frequency tolerance as described in section 5.5.

5.7. CRYSTAL AGING

Crystal aging is caused primarily by a gradual transfer of mass to or from the crystal blank and by a relaxation of stresses. Generally it is slowed down by operating the crystal at low drive level and at low temperature; however, it is most important that the crystal be kept clean. For this reason, it is essential that the hermetic seal of the crystal be preserved. Aging of cold-weld and glass enclosed crystals is significantly slower than that of crystals in solder sealed cans, since they can be kept cleaner. Glass enclosed crystals usually age up in frequency due to an apparent reduction in the mass of the quartz blank, while metal enclosed crystals age down in frequency because impurities settle on the blank.

Aging rate specifications are generally ± 0.0005 percent per month for standard military-type (MIL-type) crystals; however, it is possible to achieve aging rates as low as 1 part in 10^{11} per day for precision crystals. Ordinary crystals enclosed in cold-weld holders can be expected to age 1–5 parts in 10^8 per week after the first year. Aging is not accounted for in the overall temperature specification as discussed previously.

5.8. Q AND STIFFNESS OF CRYSTALS

The Q of ordinary or MIL-type crystals is normally not specified, but for standard units, it usually falls between 20,000 and 200,000. Precision crystals may have Q values as high as 5×10^6. Q is defined as X_L/R_1, where X_L is the reactance of L_1 at the operating frequency.

The C_0/C_1 ratio of a crystal usually is not specified. It is a measure of the stiffness of the crystal, as can be seen from equation (5-3). When the pulling characteristics of a crystal are important, it should be specified. Typical C_0/C_1 ratios may be on the order of 1000, although it is possible to achieve C_0/C_1 ratios from 125 to over 35,000.

5.9. MECHANICAL OVERTONE CRYSTALS

The AT-cut crystals may be operated on their fundamental frequency or on odd mechanical overtones, notably the third and the fifth overtones. Overtone crystals normally are used above 20 MHz. They have higher Q values, better aging rates, and are electrically stiffer than fundamental crystals of the same frequency. A tuned circuit is necessary in the oscillator to ensure operation on the proper overtone. Overtone crystals are frequently operated at series resonance. The overtone responses of a crystal should not be confused with harmonics of the fundamental frequency. They are two different phenomena. The overtone responses of a crystal are in general not exactly multiples of the fundamental frequency, although they are close. These overtone responses are depicted in Figure 5-8 which shows, in general, the various responses which may be expected in a typical AT-cut crystal. The spurious responses are discussed in section 5.10.

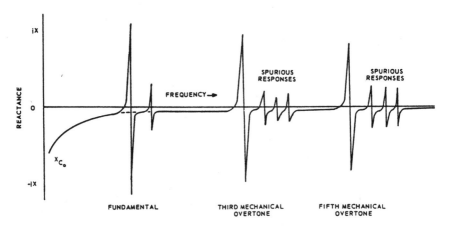

Figure 5-8. Overtone response of a quartz crystal.

As a general rule, third-overtone crystals are used from 20 to 60 MHz and fifth overtones from 60 to 125 MHz.*

5.10. SPURIOUS OR UNWANTED MODES

There are always a number of spurious responses in a quartz crystal in addition to the response of interest. This results from the fact that various modes of vibration are possible in any given quartz blank. Although the number, magnitude, and frequencies of the unwanted modes vary from crystal to crystal, an arbitrary arrangement is shown in Figure 5-8. Most of the spurious responses have a high resistance compared to the main response; however, a few low-resistance responses usually exist. They are almost always higher in frequency than the main response, and for AT-cut crystals very often fall within 200 kHz of the main response. If a spurious response has a resistance which is too low with respect to the main response, the oscillator circuit may operate on the frequency of the spurious rather than on the main response.

Generally, no problem with spurious responses is encountered using fundamental-mode AT-cut crystals. With overtone crystals, however, problems frequently are encountered. It is desirable to specify a large spurious-to-main-response resistance ratio to avoid the possibility of trouble. Practically, however, it is difficult to eliminate the unwanted responses, although several techniques are available to reduce them. With third overtone crystals, a 2-to-1 spurious ratio specification is fairly common, although often inadequate, while a 4-to-1 ratio is practical even in large production quantities. With fifth-overtone crystals, it is somewhat more difficult to make the spurious resistance high, but a 3-to-1 minimum ratio is still practical. It is often desirable to specify not only a minimum ratio but also a minimum permissible spurious resistance. This results from the fact that a larger spurious ratio is required when the crystal resistance is low. For a discussion of the spurious effects in oscillator circuits, the reader is referred to section 9.5.

---

*A considerable amount of research is being conducted in the area of surface acoustic wave resonators. These devices, which may be represented by the same equivalent circuit as the bulk wave resonators discussed in this section, show promise of extending the frequency range of crystal oscillators into the low gigahertz region. The C_0/C_1 ratio of these devices is roughly equivalent to a fifth overtone AT-cut; however, the TC is parabolic in shape.

Table 5-2. Summary of Selected MIL Crystals (compiled from MIL-STD-683D).

MIL CRYSTAL TYPE	HOLDER* TYPE	OVER-ALL FREQUENCY TOLERANCE (%)	OPERATING TEMP RANGE (°C)	LOAD COND	MODE	RATED DRIVE LEVEL (MW)
CR-18A/U	HC-6/U	±.005	−55 TO +105	32 PF	FUND	10/5 (≤10 MC / >10 MC)
CR-19A/U	HC-6/U	±.005	−55 TO +105	SERIES	FUND	10/5 (≤10 MC / >10 MC)
CR-25B/U	HC-6/U	±.01	−40 TO +85	SERIES	FUND	2.0
CR-26A/U	HC-6/U	±.002	+70 TO +80	SERIES	FUND	2.0
CR-27A/U	HC-6/U	±.002	+70 TO +80	32 PF	FUND	5/2.5 (≤10 MC / >10 MC)
CR-28A/U	HC-6/U	±.002	+70 TO +80	SERIES	FUND	5/2.5 (≤10 MC / >10 MC)
CR-32A/U	HC-6/U	±.002	+70 TO +80	SERIES	THIRD	2/1 (≤25 MC / >25 MC)
CR-35A/U	HC-6/U	±.002	+80 TO +90	SERIES	FUND	5/2.5 (≤10 MC / >10 MC)
CR-36A/U	HC-6/U	±.002	+80 TO +90	32 PF	FUND	5/2.5 (≤10 MC / >10 MC)
CR-37A/U	HC-13/U	±.02	−40 TO +70	20 PF	FUND	2.0
CR-38A/U	HC-13/U	±.012	−40 TO +70	20 PF	FUND	0.1
CR-42A/U	HC-13/U	±.003	+70 TO +80	32 PF	FUND	2.0
CR-47A/U	HC-6/U	±.002	+70 TO +80	20 PF	FUND	2.0
CR-50A/U	HC-13/U	±.012	−40 TO +70	SERIES	FUND	0.1
CR-52A/U	HC-6/U	±.005	−55 TO +105	SERIES	THIRD	4.0/2.0 (≤25 MC / >25 MC)
CR-54A/U	HC-6/U	±.005	−55 TO +105	SERIES	FIFTH	2.0
CR-55/U	HC-18/U	±.005	−55 TO +105	SERIES	THIRD	2.0
CR-56A/U	HC-18/U	±.005	−55 TO +105	SERIES	FIFTH	2.0
CR-59A/U	HC-18/U	±.002	+80 TO +90	SERIES	FIFTH	1.0
CR-60A/U	HC-18/U	±.005	−55 TO +105	SERIES	FUND	5.0
CR-61/U	HC-18/U	±.002	+80 TO +90	SERIES	THIRD	2.0/1.0 (≤25 MC / >25 MC)
CR-63B/U	HC-6/U	±.01	−40 TO +70	20 PF	FUND	2.0
CR-64/U	HC-18/U	±.005	−55 TO +105	30 PF	FUND	5.0
CR-65/U	HC-6/U	±.001	+70 TO +80	SERIES	THIRD	2.0/1.0 (≤25 MC / >25 MC)
CR-66/U	HC-6/U	±.002	−55 TO +105	30 PF	FUND	10.0/5.0 (≤10 MC / >10 MC)
CR-67/U	HC-18/U	±.0025	−55 TO +105	SERIES	THIRD	2.0
CR-68/U	HC-6/U	±.002	+70 TO +80	32 PF	FUND	5.0
CR-69A/U	HC-18/U	±.002	−55 TO +105	30 PF	FUND	5.0
CR-71/U	HC-30/U	±.00008		32 PF	FIFTH	70 UA
CR-74/U	HC-26/U	±.001	+80 TO +90	SERIES	FIFTH	1.0
CR-75/U	HC-6/U	±.001	+70 TO +80	SERIES	FIFTH	1.0
CR-76/U	HC-18/U	±.0025	−55 TO +105	SERIES	THIRD	2.0
CR-77/U	HC-25/U	±.002	−55 TO +105	SERIES	THIRD	2.0
CR-78/U	HC-25/U	±.005	−55 TO +105	30 PF	FUND	5.0
CR-79/U	HC-25/U	±.005	−55 TO +105	SERIES	FUND	5.0
CR-80/U	HC-18/U	±.003	−55 TO +105	SERIES	FIFTH	2.0
CR-81/U	HC-25/U	±.005	−55 TO +105	SERIES	THIRD	2.0
CR-82/U	HC-25/U	±.005	−55 TO +105	SERIES	FIFTH	2.0
CR-83/U	HC-25/U	±.0025	−55 TO +105	SERIES	FIFTH	2.0
CR-84/U	HC-25/U	±.002	+80 TO +90	SERIES	THIRD	2.0/1.0 (≤25 MC / >25 MC)
CR-85/U	HC-6/U	±.0025	−55 TO +105	SERIES	FUND	10/5 (≤10 MC / >10 MC)
CR-101/U	HC-35/U	±.0025	−55 TO +105	30 PF	FUND	5
CR-102/U	HC-35/U	±.0025	−55 TO +105	SERIES	FIFTH	2
CR-103/U	HC-35/U	±.0025	−55 TO +105	SERIES	THIRD	2

*FURTHER DETAILS OF THE HOLDERS ARE SHOWN IN FIGURE 5-10.

Table 5-2. (*Continued*)

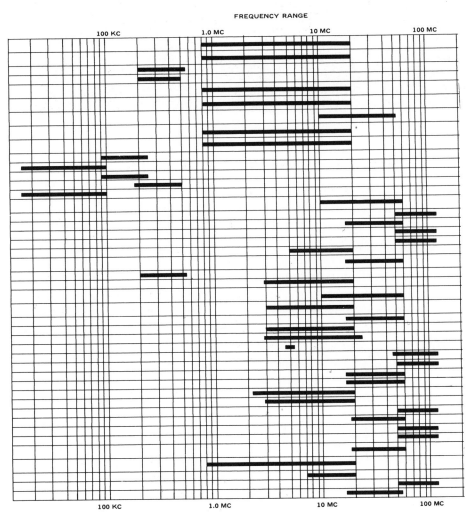

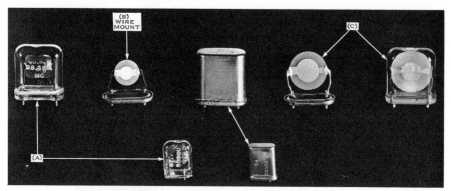

Figure 5-9. Ruggedized crystal mounts.

5.11. VIBRATION, SHOCK, AND ACCELERATION

Crystal units are available which will meet most environmental specifications. In general, vibration and shock do not cause catastrophic failures but, rather, frequency shifts and resistance changes. Frequency shifts on the order of 0.0001 percent are common, and resistance changes of 10 percent may occur. Figure 5-9 shows several ruggedized crystal mounts. The wire-mounted crystal is generally not satisfactory for severe environmental conditions, and one of the ruggedized versions must be used. (A) and (C) in Figure 5-9 generally are satisfactory for vibration up to 2000 Hz. Large crystal blanks are difficult to ruggedize and, consequently, low-frequency crystals should be avoided if severe environmental conditions will be encountered. For more specific information on environmental conditions, the reader may consult vibration specifications in MIL-C-3098.*

5.12. STANDARD MILITARY CRYSTALS

It is possible to specify a crystal to fit the needs of a particular oscillator circuit. Where possible, however, it is more desirable to use standard crystals. Table 5-2 presents a summary of selected MIL crystals, while Table 5-3 gives the maximum resistance for these units.

---

*AT-cut resonators generally show an acceleration sensitivity of about 1 ppb/g. Research is presently being conducted, however, to develop a stress-compensated crystal cut (SC) which shows promise of reducing the sensitivity by more than an order of magnitude. This also results in a reduction of the frequency overshoot during warm up due to thermal stress in the crystal blank.

TABLE 5-3. Maximum Crystal Resistance. (Compiled from
MIL-C-3098F, 24 July 1973)

CR-18A/U		CR-19A/U	
MHz	(Ω)	MHz	(Ω)
0.8 to 0.85	625	2.6+ to 3	90
0.85+ to 0.9	600	3+ to 3.4	70
0.9+ to 1	575	3.4+ to 3.75	52
1+ to 1.12	540	3.75+ to 4	45
1.12+ to 1.25	490	4+ to 5	37
1.25+ to 1.37	450	5+ to 7	25
1.37+ to 1.5	410	7+ to 10	20
1.5+ to 1.62	375	10+ to 15	18
1.62+ to 1.75	330	15+ to 20	15
1.75+ to 1.87	300	CR-25A/U	
1.87+ to 2	290	kHz	(Ω)
2+ to 2.12	270		
2.12+ to 2.25	245	200 to 225	2,500
2.25+ to 2.6	195	225+ to 265	3,000
2.6+ to 3	150	265+ to 290	3,500
3+ to 3.4	110	290+ to 330	4,000
3.4+ to 3.75	90	330+ to 370	4,500
3.75+ to 4	75	370+ to 410	5,000
4+ to 5	60	410+ to 425	5,500
5+ to 7	35	425+ to 460	6,500
7+ to 10	24	460+ to 500	7,500
10+ to 15	22	CR-26A/U	
15+ to 20	20	Same as CR-25A/U	
CR-19A/U		CR-27A/U	
MHz	(Ω)	MHz	(Ω)
0.8 to 0.85	520	0.8 to 0.85	620
0.85+ to 0.9	480	0.85+ to 0.9	600
0.9+ to 1	440	0.9+ to 1	570
1+ to 1.12	400	1+ to 1.12	540
1.12+ to 1.25	380	1.12+ to 1.25	490
1.25+ to 1.37	340	1.25+ to 1.37	450
1.37+ to 1.5	300	1.37+ to 1.5	410
1.5+ to 1.62	275	1.5+ to 1.62	370
1.62+ to 1.75	250	1.62+ to 1.75	330
1.75+ to 1.87	220	1.75+ to 1.87	300
1.87+ to 2	185	1.87+ to 2	290
2+ to 2.12	165	2+ to 2.12	270
2.12+ to 2.25	150	2.12+ to 2.25	240
2.25+ to 2.6	125	2.25+ to 2.6	190

TABLE 5-3. (Continued)

CR-27A/U		CR-50A/U	
MHz	(Ω)	kHz	(Ω)
2.6+ to 3	150	16 to 30	100,000
3+ to 3.4	110	30+ to 50	90,000
3.4+ to 3.75	90	50+ to 70	80,000
3.75 to 4	75	70+ to 90	70,000
4+ to 5	60	90+ to 100	60,000
5+ to 7	35		
7+ to 10	24	**CR-52A/U**	40 Ω
10+ to 15	22		
15+ to 20	20		

CR-28A/U		CR-54A/U	
Same as CR-19A/U		MHz	(Ω)
		50 to 90	50
CR-35A/U		90+ to 125	60
Same as CR-19A/U			

CR-36A/U		**CR-55/U**	40 Ω
Same as CR-27A/U		**CR-56A/U**	60 Ω

CR-37A/U		CR-59A/U	
kHz	(Ω)	MHz	(Ω)
90 to 170	5,000	50 to 500	50
170+ to 250	5,500	100+ to 125	60

CR-38A/U		CR-60A/U	
MHz	(Ω)	MHz	(Ω)
16 to 50	110,000	5 to 7	50
50+ to 80	100,000	7+ to 10	30
80+ to 100	90,000	10+ to 15	25

CR-42A/U		15+ to 20	20
kHz	(Ω)	**CR-61/U**	40 Ω
90 to 170	4,500		
170+ to 250	5,000		

CR-47A/U		CR-63B/U	
kHz	(Ω)	MHz	(Ω)
190 to 225	3,700	200 to 225	5,300
225+ to 275	4,200	225+ to 275	6,000
275+ to 325	4,600	275+ to 325	6,500
325+ to 375	4,900	325+ to 375	7,000
375+ to 425	5,300	375+ to 425	7,500
425+ to 475	5,600	425+ to 475	8,000
475+ to 500	6,000	475+ to 500	8,500
		500+ to 555	5,000

TABLE 5-3. (Continued)

CR-64/U	
MHz	(Ω)
2.9 to 3.75	180
3.75+ to 4.75	120
4.75+ to 6	75
6+ to 7	50
7+ to 10	30
10+ to 20	25

CR-65/U	40 Ω

CR-66/U	
MHz	(Ω)
3 to 4	60
4+ to 5	50
5+ to 7	45
7+ to 10	35
10+ to 20	25

CR-67/U	40 Ω

CR-68/U	
MHz	(Ω)
3 to 4	40
4+ to 5	35
5+ to 6	30
6+ to 7	28
7+ to 8	25
8+ to 9	23
9+ to 10	20
10+ to 15	18
15+ to 20	15

CR-69/U	
MHz	(Ω)
2.9 to 3.75	180
3.75+ to 4.75	120
4.75+ to 6	75
6+ to 7	50
7+ to 10	30
10+ to 25	25

CR-71/U	175 Ω

CR-74/U	50 Ω

CR-75/U	40 Ω

CR-76/U	40 Ω

CR-77/U	40 Ω

CR-78/U	
MHz	(Ω)
2.2 to 3.00	360
3.0+ to 3.75	180
3.75+ to 4.75	120
4.75+ to 6	75
6+ to 7	50
7+ to 10	30
10+ to 20	25

CR-79/U	
MHz	(Ω)
2.9 to 7.0	50
7+ to 10	30
10+ to 15	25
15+ to 20	20

CR-80/U	
Same as CR-54A/U	

CR-81/U	40 Ω

CR-82/U	
Same as CR-54A/U	

CR-83/U	
Same as CR-54A/U	

CR-84/U	40 Ω

CR-85/U	
Same as CR-19A/U	

CR-101/U	
7 to 10	30
10+ to 20	25

CR-102/U	60 Ω

CR-103/U	40 Ω

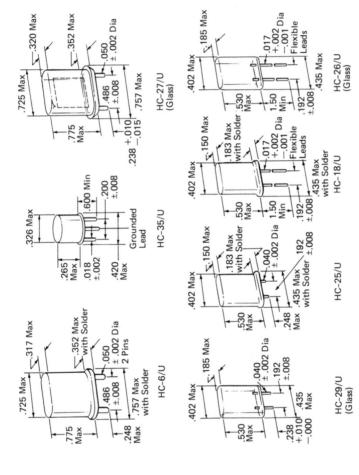

Figure 5-10. Crystal holder dimensions.

The dimensions of the more commonly used crystal holders are given in Figure 5-10. For additional information the reader may consult the latest version of MIL-C-3098. In some applications only a few parameters will be different from a standard crystal, and here the military specifications are a good basic guideline to use in writing the specification. An example of a crystal specification which refers to several MIL standards is given in Appendix L.

5.13. SPECIFICATIONS AND STANDARDS

A large number of specifications and standards are available which present very useful information on crystals and methods of measurement. Among these are IEEE Standards, EIA Standards, IEC Standards, and MIL Standards. A good listing of these is presented in reference 50, p. 494 along with information on where they may be obtained.

6
Discussion of
Transistors

The selection of a transistor type for a crystal oscillator is largely based on engineering judgment. The following factors that should be considered are discussed in this chapter:

a. Temperature requirements.
b. Maximum frequency requirements.
c. Output power requirements.
d. Input and output impedance.
e. Available power gain.
f. Interchangeability requirements.
g. Cost, availability, etc.

In addition to these, several other characteristics affect the oscillator performance to a lesser degree. A number of these are also discussed.

The transistor chosen must obviously be operable over the required temperature range. In addition, the variation in transistor characteristics with temperature must be compatible with the oscillator circuit. Roughly, the β of a transistor decreases by about 50 percent from room temperature to $-55°C$ and increases by about 50 percent from room temperature to the maximum permissible operating temperature. However, some transistors are better than others in this respect. The saturation resistance increases with temperature. At VHF frequencies, the characteristics of transistors are less dependent on temperature, since the low-frequency parameters have little influence on VHF performance.

Generally, bipolar transistors are used for crystal oscillators because of their larger transconductance at low power levels. The use of field effect devices as crystal oscillators is increasing, however, in connection with oscillators implemented with logic gates and other integrated circuits.

The cutoff frequency f_t of a transistor is some measure of the phase shift through it at the operating frequency. This phase shift is lagging and causes the oscillator frequency to decrease and to become more dependent on the transistor. Therefore, it is desirable to use a transistor with a cutoff frequency at least an order of magnitude higher than the operating frequency.

Larger output powers obviously require higher transistor dissipation. It must be remembered, however, that the allowable quartz crystal dissipation often limits the maximum power output of a stable oscillator.

For high-stability oscillators, it is desirable to minimize the effects of the transistor on the frequency. For this reason the input and output capacitances of the transistor are often swamped out by the addition of external input and output capacitors. (This is particularly convenient in the Pierce, Colpitts, and Clapp oscillator circuits.) If the input and output capacitances of the transistor are small, they can be swamped out effectively without the external capacitors becoming large enough to prevent oscillation. Therefore, it is desirable to use transistors with low input and output capacitances.

6.1. TRANSISTOR EQUIVALENT CIRCUITS

A large number of equivalent circuit representations have been used for transistors in various applications. Obviously, different representations work better or are more practical in certain applications than in others. It has been found that for crystal oscillator design, the Y-parameter representation and the hybrid π equivalent circuit are very useful. Consequently, these circuits will be reviewed briefly prior to incorporating them in the derivation of oscillator equations. The Y-parameter representation is quite versatile in that any linear device can be characterized using the approach, whether it be a single transistor or a combination of devices such as a gate or an integrated amplifier. It is often more convenient to perform measurements on a device, particularly at VHF frequencies, using S-parameters; therefore, the equations required to convert from S-parameters to Y-parameters are also included.

It is possible to perform a complete oscillator analysis based on S-parameters. Such an analysis would be most useful in the design of microwave oscillators. Since quartz crystals are generally not used beyond 150 MHz, the Y-parameter approach is more appropriate for the present work.

The hybrid π equivalent circuit of a transistor is, of course, much more closely related to the physical properties of the device. It is useful in the design of crystal oscillators because it can easily be adapted and used in a nonlinear model and thus is used in connection with analyses to predict the amplitude of oscillation. The Y-parameters can also be derived from the hybrid π equivalent circuit, and these are given in section 6.3.

6.2. Y-PARAMETER MODEL

The Y-parameter representation of a transistor or device is based on the assumption that the device is linear. This is a valid assumption during the initial buildup of oscillation and can therefore be used to predict the starting conditions for oscillation. The starting conditions are obviously an important part of the design and are studied in great detail. After the signal becomes large, the Y-parameters can still be useful if we define them as the ratios of the fundamental components of current to the fundamental components of voltage.

The Y-parameter representation of a device is shown in Figure 6-1 along with an equivalent circuit which can be used in Figure 6-2.

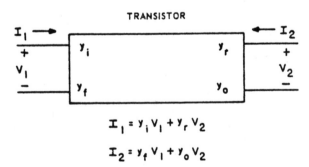

Figure 6-1. Y-Parameter representation of a transistor.

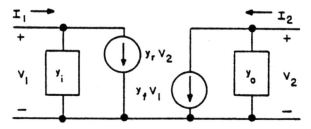

Figure 6-2. Transistor equivalent circuit using Y-parameters.

The term y_i is the input admittance with the output short-circuited:

$$y_i = \frac{I_1}{V_1}\bigg|_{V_2=0}.$$

In like manner, y_o is the short-circuit output admittance:

$$y_o = \frac{I_2}{V_2}\bigg|_{V_1=0};$$

y_f is the forward transfer admittance (also referred to as the trans-conductance):

$$y_f = \frac{I_1}{V_2}\bigg|_{V_1=0}.$$

and y_r is the reverse transfer admittance:

$$y_r = \frac{I_1}{V_2}\bigg|_{V_1=0}.$$

A subscript e, b, or c is added to the Y-parameters to indicate the circuit configuration; thus, y_{ib} is the common-base input admittance while y_{ie} is the common-emitter input admittance. If the common-emitter parameters are given on the transistor data sheet, the common-base parameters can be calculated using the following relationships:

$$y_{ib} = y_{ie} + y_{oe} + y_{fe} + y_{re} \qquad (6\text{-}1)$$

$$y_{fb} = -(y_{fe} + y_{oe}) \qquad (6\text{-}2)$$

$$y_{rb} = -(y_{re} + y_{oe}) \qquad (6-3)$$

$$y_{ob} = y_{oe} \qquad (6-4)$$

Also Y-parameters can be calculated from S-parameters as follows. If we assume a common-emitter configuration so that $y_{ie} = y_{11}$, $y_{fe} = y_{21}$, $y_{re} = y_{12}$, and $y_{oe} = y_{22}$,

$$y_{11} = \frac{(1 - S_{11})(1 + S_{22}) + S_{12}S_{21}}{Z_o[(1 + S_{22})(1 + S_{11}) - S_{12}S_{21}]} \qquad (6-5)$$

$$y_{21} = \frac{-2S_{21}/Z_o}{(1 + S_{11})(1 + S_{22}) - S_{12}S_{21}} \qquad (6-6)$$

$$y_{12} = \frac{-2S_{12}/Z_o}{(1 + S_{11})(1 + S_{22}) - S_{21}S_{12}} \qquad (6-7)$$

$$y_{22} = \frac{(1 + S_{11})(1 - S_{22}) + S_{21}S_{12}}{Z_o[(1 + S_{11})(1 + S_{22}) - S_{12}S_{21}]} \qquad (6-8)$$

For purposes of oscillator analysis, it is often convenient to break the Y-parameters into their real and imaginary parts (as listed on transistor data sheets). Therefore, the following standard designations will be used

$$y_i = g_i + jb_i$$

$$y_o = g_o + jb_o$$

$$y_f = g_f + jb_f$$

$$y_r = g_r + jb_r$$

Physically,

$$g_i = \frac{1}{R_{in}}, \quad b_i = \omega C_{in}, \quad g_o = \frac{1}{R_{out}}, \quad \text{and} \quad b_o = \omega C_{out}.$$

$$(6-9)$$

6.3. HYBRID π EQUIVALENT CIRCUIT

The hybrid π equivalent circuit of a transistor is shown in Figure 6-3. Unfortunately, at all but low frequency, where capacitances can be neglected, the application of this circuit to oscillators leads

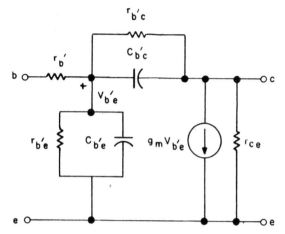

Figure 6-3. Hybrid π (common-emitter) equivalent circuit.

to complicated equations that are difficult to solve. Pritchard[39] has made certain simplifications to the equivalent circuit of Figure 6-3. Figure 6-4 shows this approximate high-frequency equivalent circuit for the common-emitter configuration.

The resulting Y-parameters are

$$y_{ie} = \frac{1}{r_{b'} - j(\omega_t/\omega)r_{e'}}$$
(6-10)

$$y_{re} = \frac{-\omega_t r_{c'} C_c}{r_{b'} - j(\omega_t/\omega)r_{e'}}$$
(6-11)

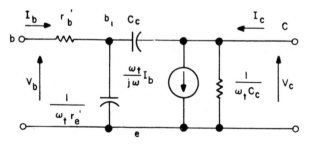

Figure 6-4. Approximate high-frequency equivalent circuit of a junction transistor.

$$y_{fe} = \frac{-j(\omega_t/\omega)}{r_{b'} - j(\omega_t/\omega)r_{e'}} \qquad (6\text{-}12)$$

$$y_{oe} = \frac{\omega_t C_c(r_{b'} + r_{e'}) + j\omega C_c(r_{b'})}{r_{b'} - j(\omega_t/\omega)r_{e'}} \qquad (6\text{-}13)$$

These equations establish the approximate correlation between internal parameters and terminal parameters. Since most of these expressions are rather involved, it is simpler to express circuit design equations in terms of the terminal parameters.[26]

6.4. NONLINEAR MODELS

In contrast to the case of small-signal properties of the transistor, very few results predicting the large-signal behavior are available in the literature. As a result, the engineer is generally unfamiliar with this aspect of transistor behavior.

As previously stated, the small-signal behavior holds only for base-to-emitter signal voltages up to about 10 mV. It is possible, however, to define equivalent linear properties at higher signal levels. This is done by forming the ratios of the fundamental components of voltage to the fundamental components of current. The ratios, of course, change with amplitude. It is the purpose of this section to present formulas which will enable the engineer to predict the large-signal properties of a transistor knowing the small-signal values and the signal amplitude. The predictions are accomplished using the hybrid π equivalent circuit. Two types of analysis are made; the first, in section 6.4.1, is valid for the intrinsic transistor, neglecting the base spreading resistance, and is useful for most crystal oscillator applications below 10–20 MHz. In some low-noise oscillator applications, however, it has been found desirable to use emitter degeneration to reduce $1/f$ noise. Therefore a second analysis, given in section 6.4.2, is presented to allow prediction of the amplitude of oscillation when degeneration is used. The results generally follow the same form, although the mathematics used in deriving them is considerably different.

6.4.1. Intrinsic Transistor Model*

The large-signal analysis of transistor parameters presented in this paragraph is based on the hybrid π equivalent circuit shown in Figure 6-5. The various elements of this circuit are the same as those presented in Figure 6-3, although they are represented differently in some cases. It should be understood that this circuit is only an approximate equivalent circuit of the transistor and represents some rather serious simplifications of the actual device.

The resistor $r_{bb'}$ is the base-spreading resistance and is neglected in the analysis.

The emitter resistance r_e is composed of intrinsic and extrinsic parts. The intrinsic part usually accounts for the largest portion of the resistance and, for small-signal conditions, is given by

$$r_{e0} = \frac{KT\lambda}{qI_e};$$

(6-14)

at 27°C and $\lambda = 1$

$$r_{e0} = \frac{26}{I_e} \quad \text{with } I_e \text{ in mA.}$$

(6-15)

Here the subscript 0 refers to a small-signal value. As will be shown later, r_e varies considerably with signal level.

The base diffusion capacitance C_{De} is given by k/r_e and generally accounts for most of the transistor input capacitance in the active region.

The diffusion capacitance is related to the gain–bandwidth product, so that

$$k \doteq \frac{1}{2\pi f_t}.$$

(6-16)

C_{Te} is base-to-emitter transition capacitance (junction capacitance), which depends on the size of the base-to-emitter junction. It

---

*The application of these results to crystal oscillators is essentially parallel to the results discussed in connection with LC oscillators by Holford in Mullard Technical Communications, see reference 21.

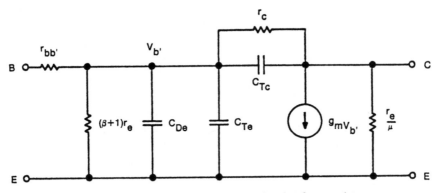

Figure 6-5. Hybrid π equivalent circuit of a transistor.

generally varies as the square root of the base-to-emitter voltage; thus:

$$C_{Te} = \frac{K_{Te}}{(V - V_{be})^{1/2}} \qquad (6\text{-}17)$$

The base-to-collector transition (junction) capacitance C_{Tc} is dependent on the junction area and on the grading of the junction. It varies as some power of the applied voltage; thus:

$$C_{Tc} = \frac{K_{Tc}}{[V_{cb} + V]^{\alpha}} \qquad (6\text{-}18)$$

where V is the contact potential and α is a function of the grading of the junction; α is usually between 0.5 and 0.1.

Transconductance is given approximately by $1/r_e$, and the feedback factor is given by μ which is considered to be a constant for this analysis. It is shown in Appendix G that r_e increases with signal level and is given by

$$\frac{r_e}{r_{e0}} = \frac{\left(\dfrac{Eq}{\lambda KT}\right) I_0 \left(\dfrac{Eq}{\lambda KT}\right)}{2I_1 \left(\dfrac{Eq}{\lambda KT}\right)} \qquad (6\text{-}19)$$

where r_{e0} is the small-signal emitter resistance given by equation (6-14). E is the peak value of the base-to-emitter signal voltage assumed to be given by $E \cos \omega t$. For purposes of this analysis, it is convenient to define the voltages in terms of $q/\lambda KT$ units; thus, we let

$$V = \frac{qE}{\lambda KT}. \tag{6-20}$$

At 27°C and $\lambda = 1$,

$$V = \frac{E}{26} \tag{6-21}$$

when E is in millivolts. Thus, we have

$$\frac{r_e}{r_{e0}} = \frac{VI_0(V)}{2I_1(V)} \tag{6-22}$$

$I_0(V)$ and $I_1(V)$ are hyperbolic Bessel functions of the first kind of orders zero and one, respectively. Equation (6-22) is plotted in Figure 6-6 along with its reciprocal.

From the transistor equivalent circuit, we see that most of the parameters are either proportional to or inversely proportional to r_e. Thus knowing how r_e behaves with signal voltage, we know also how the input and output resistance, the transconductance, and the input capacitance behave.

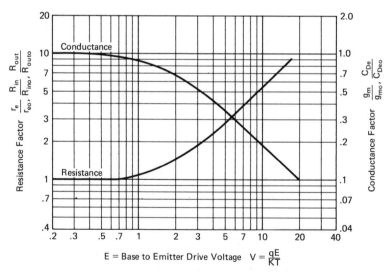

Figure 6-6. Transistor parameters versus signal voltage.

It should be pointed out that the analysis requires the small-signal parameters to be calculated at the final emitter current with signal applied. In general, this current is higher than the value with no signal. An equation is derived in appendix G which allows the final emitter current to be predicted. This is done by computing the change in the dc base-to-emitter voltage resulting from the signal. It is then only necessary to allow for this decrease in base-to-emitter voltage, V bias, when computing the values of the bias resistors. This equation is given by

$$\frac{qV_{bias}}{\lambda KT} = \ln I_0(V) \tag{6-23}$$

and is plotted in Figure 6-7.

It is possible to predict the amount of fundamental and harmonic

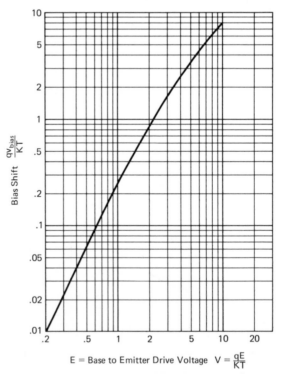

Figure 6-7. Base-to-emitter bias shift versus applied signal voltage.

signal present in the collector current when a sinusoidal signal is applied to the base. The equation predicting this behavior (see Appendix G), neglecting feedback, is given by:

$$i_c = 2I_e(\text{mean}) \left[\frac{I_1(V)}{I_0(V)} \cos \omega t + \frac{I_2(V)}{I_0(V)} \right.$$

$$\left. \cos 2\omega t + \frac{I_3(V)}{I_0(V)} \cos 3\omega t + \cdots \right] \quad (6\text{-}24)$$

This equation is plotted in Figure 6-8 along with the equation giving the peak collector current, which is

$$i_c(\text{peak}) = I_e(\text{mean}) \, e^v/I_0(V) \quad (6\text{-}25)$$

This graph may be used to determine the efficiency of transistors or oscillators used as frequency multipliers. It is interesting to note

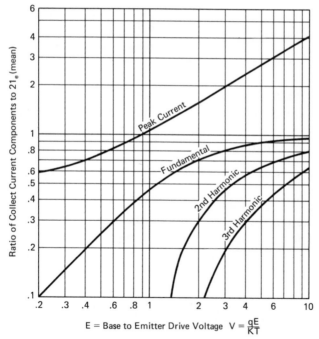

Figure 6-8. Collector current components versus applied signal voltage.

that the fundamental component of collector current approaches $2I_e(\text{mean})$ for large drive voltages.

The results presented in this paragraph are derived in Appendix G, as previously indicated. The appendix also contains a comparison of the predicted and measured results for a typical oscillator transistor using different values of emitter current.

Depending on the degree to which the assumptions made are valid, the analysis may on occasions yield a large error in the absolute value of the amplitude. In these cases the trend shown by the analysis may still be useful, however.

6.4.2. Nonlinear Model with Emitter Degeneration*

Basically the same types of curve have been derived for the case where the extrinsic emitter resistance is predominant. For this case, the equivalent circuit assumes the form shown in Figure 6-9, where R_f is the emitter degeneration resistor. Here the feedback term μ is assumed to be negligible. Also for this analysis $(r_{bb'} + R_f)_0$ is assumed to be negligible compared to $\beta(R_f + r_e)_0$. Under these conditions, it is shown in Appendix H that

$$\frac{(R_f + r_e)}{(R_f + r_e)_0} = \frac{\pi}{\theta - \frac{1}{2}\sin 2\theta} \qquad (6\text{-}26)$$

Here again the subscript 0 refers to a small-signal value. θ is one-half the effective conduction angle in radians. The conduction angle can be computed in terms of the signal voltage and the mean emitter current by the equation:

$$\sin\theta - \theta\cos\theta = \frac{I_e(\text{mean})\,(r_e + R_f)_0\pi}{E} \qquad (6\text{-}27)$$

where E is the peak value of the signal voltage applied between the base and ground, assumed to be of the form $E\sin\omega t$. Equations (6-26) and (6-27) can be plotted parametrically and are given in Figure 6-10 along with the reciprocal of equation (6-26). It should be noted that if R_f is zero, the term

$$\frac{E}{I_e(\text{mean})\,(r_e + R_f)_0} = \frac{E}{I_e(\text{mean})\,r_{e0}} = \frac{Eq}{KT\lambda}$$

*The application of these results to crystal oscillators is essentially parallel to the results discussed in connection with LC oscillators by Holford in Mullard Technical Communications, see reference 22.

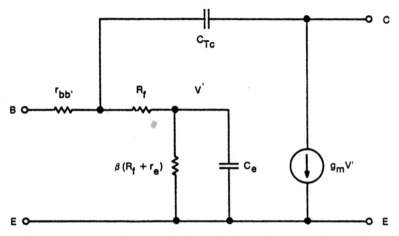

Figure 6-9. Approximate equivalent circuit with emitter degeneration.

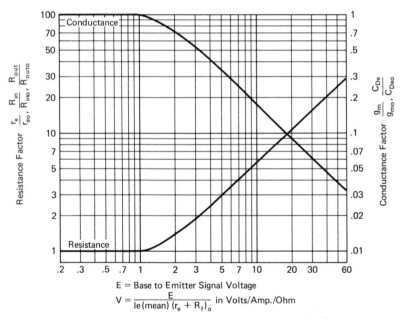

E = Base to Emitter Signal Voltage

$$V = \frac{E}{Ie \, (mean) \, (r_e + R_f)_o} \text{ in Volts/Amp./Ohm}$$

Figure 6-10. Transistor parameters versus signal voltage.

as before. The curves are not identical, however, because the latter analysis acquires a considerable error if R_f is not larger than r_e.

As before, it is possible to compute the reduction in base-to-emitter voltage due to the presence of signal, which in this case is given by

$$\frac{V_{bias}}{I_e(\text{mean})\,(r_e + R_f)_0} = \frac{\pi}{\tan\theta - \theta} + 1. \tag{6-28}$$

This equation is plotted in Figure 6-11.

It is also possible to predict the harmonic content of the collector current, the ac value of which is given by

$$i_c = I_e(\text{mean})\left\{ \frac{[\theta - \frac{1}{2}\sin 2\theta]}{[\sin\theta - \theta\cos\theta]}\sin\omega t \right.$$

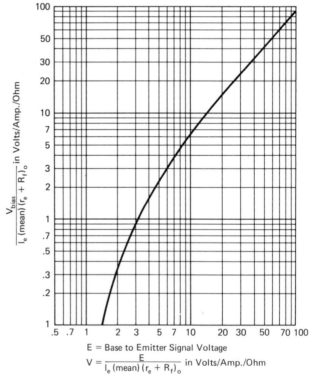

E = Base to Emitter Signal Voltage

$$V = \frac{E}{I_e(\text{mean})\,(r_e + R_f)_0} \text{ in Volts/Amp./Ohm}$$

Figure 6-11. Base-to-emitter bias shift versus applied signal voltage.

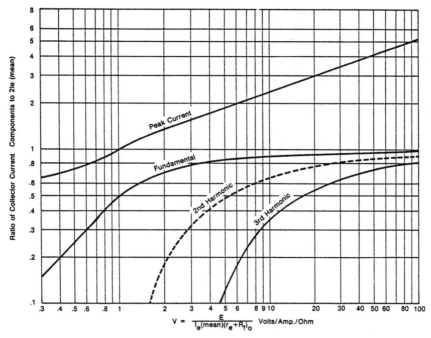

Figure 6-12. Collector current components versus applied signal voltage.

$$+ \frac{[\cos \theta \sin 2\theta - \sin \theta - \frac{1}{3} \sin 3\theta]}{[\theta \sin \theta - \theta \cos \theta]} \cos 2\omega t$$

$$\left. + \frac{[\frac{2}{3} \cos \theta \sin 3\theta - \frac{1}{2} \sin 2\theta - \frac{1}{4} \sin 4\theta]}{[\sin \theta - \theta \cos \theta]} \sin 3\omega t + \cdots \right\} \quad (6\text{-}29)$$

This equation is plotted in Figure 6-12 along with the peak collector current, which is given by the equation

$$i_c(\text{peak}) = I_e(\text{mean}) \frac{\pi(1 - \cos \theta)}{\sin \theta - \theta \cos \theta} \quad (6\text{-}30)$$

The derivation of the results presented in this section is made in Appendix H.

7

Oscillator Circuits

This chapter treats a number of crystal oscillator configurations in detail. Design information as well as practical schematic diagrams for the recommended circuits are presented. In addition, the characteristics of the various recommended oscillator types are summarized and compared in table 7-1.

7.1. PIERCE, COLPITTS, AND CLAPP OSCILLATORS

The Pierce, Colpitts, and Clapp oscillators are actually the same circuit but with the ground point at a different location. Figure 7-1 shows the basic ac schematic diagram.

In the Pierce oscillator, the ac ground is at the emitter; in the Colpitts, at the collector; and in the Clapp, at the base. In a practical circuit, the stray capacitances and biasing resistors shunt different elements for each of the three configurations, making the circuits perform somewhat differently. Each of the circuits can be made to cover a broad range of crystal frequencies. These circuits are among the most noncritical of crystal oscillators and the permissible component tolerances are generally more than adequate. The output power is only moderate, however. Of the three possible configurations, the Pierce is the most desirable, electrically. This results from the stray capacitances appearing across capacitors C_1 and C_2, which generally are quite large. In the Clapp and Colpitts configurations, a good deal of the stray capacitance appears across the crystal, limiting the high-frequency application to about 30 MHz.

In the Pierce and Clapp oscillators, the base-biasing resistors are across large capacitors and thus do not affect the performance of the circuit. In the Colpitts configuration, however, the biasing resistors are across the crystal and degrade performance at lower frequencies (below about 3 MHz). The Colpitts configuration also is more sus-

Table 7-1. Recommended Crystal Oscillator Types.

Oscillator Type	Recommended Frequency Range	Relative Frequency Stability	Power Output	Waveform	Ability to Operate Properly When Circuit Stray Capacitance and Inductance Are Large	Ability to Operate Over a Band of Frequencies Without Retuning	Ease of Design	Remarks	Paragraph
Gate	16 kHz to 20 MHz	Low	Moderate	Square wave	Good	High	Moderate	Recommended for logic level output in low-stability applications.	7.6
Pierce	100 kHz to 20 MHz	High	Moderate	Poor at low freq, fair to good above 3 MHz	Very good	High	Simple	Recommended unless one side of crystal must be grounded.	7.2
Colpitts	1 to 20 MHz	Moderate	Moderate	Fair to good	Good	High	Moderate	Generally inferior to Pierce and Clapp. Recommended if Pierce and Clapp cannot be used.	7.3
Clapp	2 to 20 MHz	Moderate to high	Moderate	Fair to good	Good	High	Moderate	Generally inferior to Pierce. Recommended if one side of crystal must be grounded. Should not be used with low supply voltages.	7.4
Impedance inverting Pierce	20 to 75 MHz	High	Low	Good	Fair	Low	Difficult	Recommended if large stray inductances cannot be eliminated from crystal switch.	7.2.9 thru 7.2.11
Grounded base	20 to 150 MHz	Moderate	High	Good	Poor	Low	Moderate	Recommended if stray inductance and capacitance can be kept low.	7.5

ceptible to squegging. These problems can be overcome using a field effect transistor for lower frequencies, since very large gate-biasing resistors then can be used.

The Clapp oscillator has a unique disadvantage in that free-running oscillations can occur if a choke is used to supply the dc voltage to the collector. The problem is best solved by putting a fairly large resistor in series with the choke or by using a resistor alone. The resistance must be kept large, however, since it shunts the crystal. For this reason, the Clapp oscillator is not desirable for use with low supply voltages.

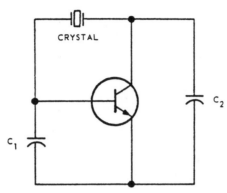

Figure 7.1. Pierce, Colpitts, and Clapp oscillators: basic ac schematic diagram.

Of the three possible configurations, the Pierce oscillator is generally the simplest and the Colpitts the most difficult to design. The Pierce oscillator has the disadvantage that one side of the crystal cannot be grounded, often making it undesirable for use with crystal switches.

The frequency stability of the Pierce oscillator is generally in the range from 0.0002 to 0.0005 percent worse than the stability of the crystal alone. The Clapp oscillator is slightly inferior to the Pierce and the Colpitts is slightly inferior to the Clapp in this respect. If no adjustment is provided to put the crystal exactly on frequency, additional frequency errors will be present as a result of differences in transistors, components, and crystal resistance.

7.2. PIERCE OSCILLATOR

7.2.1. Small-Signal Analysis

The general oscillator theory presented in Chapter 2 can be applied conveniently to the Pierce oscillator. Specifically, the conditions of oscillation are fulfilled in the following way. Referring to Figure 7-2, the basic phase shift network is composed of C_1, C_2, and the crystal, which looks inductive. Capacitors C_1 and C_2 are normally so large that they effectively swamp out the transistor output and input impedances. If this is the case, and if the effective resistance of the crystal is low, then the following explanation is applicable. The crystal looks inductive and is series resonant with capacitors C_1 and

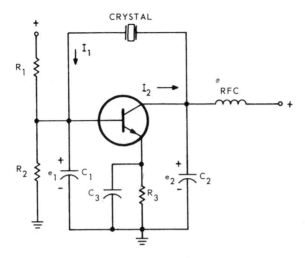

Figure 7-2. Pierce oscillator circuit: schematic diagram.

C_2. The frequency of oscillation automatically adjusts itself so that this is true. Therefore, the combination of the crystal and C_1 alone has a net inductive reactance at the operating frequency.

Consequently, current I_1 lags voltage e_2 by 90 degrees. Voltage e_1 being developed across capacitor C_1 lags current I_1 by 90 degrees, making it 180 degrees behind collector voltage e_2. Since the combination of C_1 and the crystal is resonant with C_2, the collector looks into a resistive load.

The phase shift through the transistor is 180 degrees and the total phase shift around the loop is 360 degrees.

The condition of a loop gain of unity can be found in the following manner. It can be shown (see Appendix E) that the ratio of the voltages

$$\frac{e_1}{e_2} \doteq - \left(\frac{C_2}{C_1}\right). \tag{7-1}$$

Putting this in terms of reactances gives

$$\frac{e_1}{e_2} \doteq - \left(\frac{X_1}{X_2}\right). \tag{7-2}$$

For oscillation to take place, the transistor gain A must be such that

$$A \left(\frac{e_1}{e_2} \right) \geqslant 1. \tag{7-3}$$

The transistor voltage gain is approximately given by

$$A = -g_{fe} Z_L \tag{7-4}$$

where g_{fe} is the transconductance of the transistor and Z_L is the load seen by the collector. It can be shown (see Appendix E) that this load is given by

$$Z_L \doteq \frac{X_2^2}{R_e},$$

where

$$X_2 = - \frac{1}{\omega C_2}. \tag{7-5}$$

Then

$$A = \frac{-g_{fe}(X_2)^2}{R_e} \tag{7-6}$$

and substituting this into equation 7-3 gives

$$-\left[\frac{g_{fe}(X_2)^2}{R_e} \right] \left[\frac{e_1}{e_2} \right] \geqslant 1 \tag{7-7}$$

but since

$$\frac{e_1}{e_2} = - \left(\frac{X_1}{X_2} \right), \quad \frac{g_{fe}(X_2)^2}{R_e} \left(\frac{X_1}{X_2} \right) \geqslant 1 \tag{7-8}$$

or

$$g_{fe} X_1 X_2 \geqslant R_e \tag{7-9}$$

The loop gain is then

$$\left(\frac{g_{fe} X_1 X_2}{R_e} \right) \tag{7-10}$$

and must be greater than unity. To a first approximation we may use $g_{fe} = 0.04 I_e$, where g_{fe} is in mhos and I_e is in milliamperes.

This explanation is somewhat idealized because of the assumptions made. A rigorous analysis of the Pierce oscillator is given in Appendix B. The results are.

$$g_{fe} X_1 X_2 \geqslant R_e + K_1 \qquad \text{(gain equation)} \qquad (7\text{-}11)$$

$$X_1 + X_2 + X_e = 0 + K_2 \qquad \text{(phase shift equation)} \qquad (7\text{-}12)$$

where

g_{fe} = real part of the forward transfer admittance, sometimes referred to as the transconductance.
$X_1 = -1/\omega C_1$,
$X_2 = -1/\omega C_2$,
R_e = effective crystal resistance, and
X_e = crystal reactance.

K_1 and K_2 are corrective terms which are negligible if the previous assumptions are fulfilled. They are as follows:

$$K_1 = -X_1 (X_2 + X_e) g_{ie} - X_2 (X_1 + X_e) g_{oe}$$

$$- R_e X_1 X_2 [g_{ie} g_{oe} + b_{fe} b_{re}] - b_{re} g_{fe} X_1 X_2 X_e \qquad (7\text{-}13)$$

$$K_2 = b_{fe} X_1 X_2 + X_1 X_2 X_e g_{ie} g_{oe} - R_e [X_1 g_{ie} + X_2 g_{oe}]$$

$$+ b_{re} X_1 X_2 + b_{re} b_{fe} X_1 X_2 X_e - b_{re} g_{fe} X_1 X_2 R_e. \qquad (7\text{-}14)$$

For definitions of the Y-parameters, refer to Chapter 6. The input and output short-circuit capacitances of the transistor may be accounted for by including them in C_1 and C_2, respectively. If the oscillator is loaded, the load must be included in the output admittance of the transistor; that is, g_{oe} in the equations should be $[g_{oe}$ of the transistor $+ (1/R_L)]$, where R_L is the load.

In the case of field effect transistors at low frequencies, $y_{fe} = g_m$, $y_{ie} = 0$, $y_{oe} = 1/r_d$, $y_{re} = 0$, and the equations simplify as follows:

$$g_m X_1 X_2 \geqslant R_e - \frac{X_2 (X_1 + X_e)}{r_d} \qquad (7\text{-}15)$$

$$X_1 + X_2 + X_e = - \frac{R_e X_2}{r_d} \qquad (7\text{-}16)$$

Equations (7-11) and (7-12) can be used best by successive approximation. The first values of X_1 and X_2 may be calculated assuming that $K_1 = K_2 = 0$. These values of X_1 and X_2 then are substituted into the total equations. It should be obvious then which terms are negligible, and they are eliminated. The equations then are rewritten using the significant terms. Usually the remaining equation will be only moderately complicated.

In some cases the Y-parameters of the transistor may not be known or, from other considerations, the reader may elect to use a purely experimental approach in designing a Pierce oscillator. For such an approach, the following guidelines can be given. In general, C_1 and C_2 should be as large as possible but still allow the circuit to oscillate with two to three times the maximum permissible crystal resistance.* This usually results in the crystal reactance (X_e) being quite small. A trimmer capacitor is then placed in series with the crystal to bring it on frequency. (X_e in the equation then represents the reactance of the crystal in series with the trimmer.)

In the higher portion of the frequency range, C_1 and C_2 in series may become less than the desired crystal load capacitance for minimum gain requirements. The crystal then may be put on frequency by placing a variable inductor in series with it. This may be undesirable, however, since it can lead to free-running oscillations. A better solution may be to let $C_1 = C_2 = 2C_L$. This condition leads to a maximum $X_1 X_2$ product for a given load capacitance C_L. If this does not allow sufficient gain, then a small inductor in series with the crystal must be used.

Regardless of the frequency, it is desirable to make

$$|X_1| << \frac{1}{g_{ie}}, \tag{7-17}$$

and

$$|X_2| << \frac{1}{g_{oe}}. \tag{7-18}$$

This minimizes the effects of transistor input and output admittances.

For a highly stable oscillator, the loading should be light. A good

---

*This can be determined by adding resistance in series with the crystal until oscillation will not occur.

rule to follow is to tap considerably farther down than for optimum matching. An example of this is shown in Figure 7-3, (see section 7.2.3 below), where the ratio of C_4/C_2 is larger than that required for maximum output.

Best stability occurs if C_1 and C_2 are as large as possible, because they swamp out any change in transistor input or output capacitance.

To a first approximation, it can be shown that maximum stability results if

$$\frac{C_2}{C_1} = \left[\frac{dC_{\text{out}}/dT}{dC_{\text{in}}/dT}\right]^{1/2} \tag{7-19}$$

where dC_{out}/dT is the change in collector-to-emitter capacity with respect to the parameter being studied, and dC_{in}/dT is the variation in base-to-emitter capacity with respect to the same parameter. The ratio C_2/C_1 usually is determined by other factors, however, such as crystal load capacitance, output voltage, or crystal drive. Since the load on the collector is $Z_L \doteq X_2^2/R_e$, larger outputs usually are obtained if X_2 is made large (small C_2).

It should be remembered that the equations derived using the Y-parameters are based on a linear analysis and predict starting conditions only. They give no indication concerning the final amplitude or of the crystal drive level. A method for determining the steady-state value of drive level as well as the output voltage is discussed in the following paragraphs.

7.2.2. Large-Signal Analysis

If the starting conditions for oscillation are satisfied, the amplitude of oscillation continues to grow until nonlinear effects reduce the effective loop gain to unity. If the starting conditions are satisfied at more than one frequency, oscillations begin at both frequencies and the one reaching the saturation amplitude first causes the other to die out. Normally, multiple or spurious oscillation is not a problem and will not be considered at this time.

In a transistor oscillator, the predominant nonlinearity occurs because the base-to-emitter junction is cut off during part of the cycle. As discussed in section 3.2 of Chapter 3, under certain conditions of biasing, collector saturation may also occur. Generally collector saturation tends to increase the dependence of the oscillator on the supply voltage and is therefore avoided.

Assuming that collector saturation does not occur, the amplitude of oscillation can be determined by applying the nonlinear results of section 6.4 of Chapter 6. To accomplish this, the small-signal analysis of section 7.2.1 (above) is first completed.

The small-signal transconductance required for oscillation is found from equation (7-11) and is:

$$g_m = \frac{R_e + K_1}{X_1 X_2} \tag{7-20}$$

If the actual small-signal transconductance of the transistor is given by g_{m0}, then the excess small-signal loop gain is g_{m0}/g_m. This value is used to enter the graph of Figure 6-6 and determines the value of V which will be required to reduce the loop gain to unity. The value V is the normalized base-to-emitter voltage. The actual voltage is $E \cos \omega t$ where $E = VKT/q$. As noted earlier $KT/q \doteq 26$ mV at room temperature. Then using equation (7-1), the approximate collector voltage is found to be

$$e_2 = -e_1 \left(\frac{C_1}{C_2} \right) \tag{7-21}$$

where $e_1 = E/\sqrt{2}$ volts rms.

The crystal drive level is then given by

$$P_c = \frac{e_2^2 R_e}{X_2^2} \tag{7-22}$$

and should not exceed the manufacturer's specification.

The graph of Figure 6-6 can also be used to determine the increase in the effective input and output impedances of the transistor and the reduction in input capacitance:

$$g_{ie} \doteq \frac{g_{ieo}}{R_{in}/R_{ino}} \tag{7-23}$$

The transistor input capacitance was lumped into C_1; therefore it will also decrease with amplitude, causing a slight increase in frequency. If desired, the new values of g_{ie}, g_{oe}, and C_1 can be used in calculating the K_1 used in equation (7-20).

The graphs of Chapter 6 also provide an indication of the bias shift resulting from oscillation. This is shown in Figure 6-7. The value of V found earlier is used to determine the reduction in base-to-emitter voltage. An appropriate adjustment can then be made to the bias network to establish the desired steady-state value of emitter current.

Finally, the graph of Figure 6-8 can be used to determine the harmonic content of the collector current. Since the collector feeds the input of the π network containing the crystal and since for practical purposes the crystal is an open circuit at the harmonic frequencies, the collector voltage waveform can be determined by multiplying the appropriate harmonic current by the reactance of C_2 at that frequency.

The impedance seen by the collector at the fundamental frequency is approximately given by

$$Z = \frac{X_2^2}{R_e}. \tag{7-24}$$

Therefore the ratio of harmonic voltage to fundamental voltage is given by

$$\frac{e_n}{e_{\text{fund}}} = \frac{i_n(X_2/n)}{i_{\text{fund}}(X_2^2/R_e)} \tag{7-25}$$

$$\frac{e_n}{e_{\text{fund}}} = \frac{i_n R_e}{i_{\text{fund}} n X_2} \tag{7-26}$$

Where n is the harmonic number, i_n and i_{fund} can be read from the graph of Figure 6-8 knowing V, or from equation (6-24) by taking i_n/i_{fund} as the ratio of hyperbolic Bessel functions $I_n(V)/I_1(V)$.

If the oscillator is loaded heavily, Z must be appropriately adjusted and the fundamental component of the waveform will be reduced, compared to the harmonics.

Since the base voltage of the transistor is established by the crystal current through C_1, it has the best waveform.

In some applications a tank circuit tuned to a harmonic of the crystal frequency is inserted in the collector circuit to produce a frequency multiplier. This procedure generally works quite well, and the graph of Figure 6-8 can be used to estimate the harmonic output power.

In some applications of frequency standards, it is desirable to use emitter degeneration on the transistor, for example to reduce the $1/f$ noise. In these applications the graphs of Figures 6-6 through 6-8 do not apply. A piecewise linear analysis has been included, however, which gives good results if sufficient feedback is used.

This analysis assumes that the transistor, with feedback, is linear during that part of the cycle when it conducts and is completely shut off during the remainder of the cycle. The results are presented in Figures 6-10, 6-11, and 6-12. These graphs can be used in place of Figures 6-6, 6-7, and 6-8 in the previous description. The actual base-to-emitter voltage is found to be

$$E = VI_e(\text{mean}) (R_f + r_e)_0 \qquad (7\text{-}27)$$

where R_f is the emitter degeneration resistor, r_e is the intrinsic emitter resistance, $I_e(\text{mean})$ is the steady-state emitter current, and E is the peak base-to-ground voltage in volts.

In general the nonlinear analyses are not as accurate as the analyses we are accustomed to seeing based on linear models in small-signal applications. The results are, however, quite useful in the initial design of oscillator circuits and give a reasonable approximation when the cutoff frequency of the transistor is more than 10 or 20 times the operating frequency.

7.2.3. 1- to 3-MHz Pierce Oscillator*

Typical performance characteristics for the 1- to 3-MHz Pierce oscillator shown in Figure 7-3 are given below.

a. Crystal: CR-18/U or similar.
b. Load capacitance: 32 pF.
c. Drive level: 1–6 mW, depending on frequency and crystal resistance.
d. Factors affecting frequency stability are as follows:
   1. Temperature coefficient of crystal: ±0.005 percent from −55°C to +105°C.
   2. Aging rate of crystal: See section 5.7.

---

*A number of typical oscillator circuits are presented in this chapter. The performance indicated is that observed on a single circuit and may vary considerably with layout and components used. The circuits are listed as starting points only and should not be used without optimization and thorough testing in the configuration actually used. (See Chap. 8.)

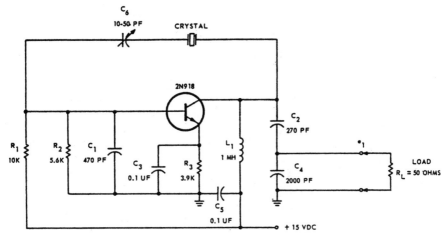

Figure 7-3. 1- to 3-MHz Pierce oscillator: schematic diagram.

3. Temperature stability of oscillator circuitry: 1.5 parts in $10^9/°C$ at 1 MHz, and 1.6 parts in $10^8/°C$ at 3 MHz.
4. Voltage coefficient of oscillator: 3 parts in 10^7 at 1 MHz, and 1 part in 10^7 at 3 MHz for a 10-percent supply voltage change.
e. Output: For R_L = 50 Ω, e_1 is approximately 0.25–0.40 V, depending on frequency and crystal resistance. Waveform at 1 MHz is poor.
f. Permissible load: $50 \leqslant R_L \leqslant \infty$.
g. Power input: 40 mW.

NOTE. Although the circuit will oscillate with any standard crystal in the 1- to-3-MHz range, some adjustment of C_6 is necessary over the frequency range to put crystals exactly on frequency.

7.2.4. 1- to 10-MHz Pierce Oscillator

Typical performance characteristics for the 1- to 10-MHz Pierce oscillator shown in Figure 7-4 are given below.

a. Crystal: CR-18/U or similar.
b. Load capacitance: 32 pF.
c. Drive level: 0.75–4 mW, depending on the frequency and crystal resistance.
d. Factors affecting frequency stability are as follows:

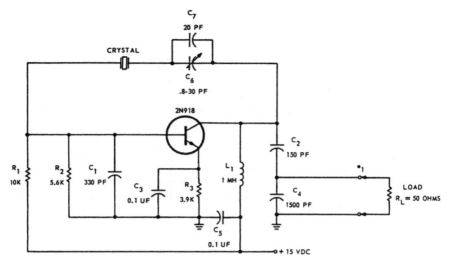

Figure 7-4. 1- to 10-MHz Pierce oscillator circuit: schematic diagram.

1. Temperature coefficient of crystal: ±0.005 percent from −55°C to +105°C.
2. Aging rate of crystal: See section 5.7.
3. Temperature stability of oscillator circuitry: 3 parts in 10^8/°C at 1 MHz, 2.2 parts in 10^8/°C at 5 MHz, and 1.5 parts in 10^8/°C at 10 MHz.
4. Voltage coefficient of oscillator: 1.5 parts in 10^6 at 1 MHz, 2.5 parts in 10^7 at 3 MHz, and 2.3 parts in 10^7 at 10 MHz for a 10-percent change in supply voltage.
e. Output: For R_L = 50 Ω, e_1 is approximately 0.02 V at 10 MHz, and 0.35 V at 3 MHz and 1 MHz. e_1 varies considerably with crystal resistance. The waveform below 3 MHz is poor. Distortion at 1 MHz is 30 percent; at 5 MHz, it is 6 percent.
f. Permissible load: 50 ⩽ R_L ⩽ ∞.
g. Input power: 35 mW.

NOTE. Although the circuit will oscillate with any standard crystal in the 1- to 10-MHz range, some adjustment of C_6 is necessary over the frequency range to put crystals exactly on frequency.

7.2.5. 10- to 20-MHz Pierce Oscillator

Typical performance characteristics for the 10- to 20-MHz Pierce oscillator shown in Figure 7-5 are given below.

  a. Crystal: CR-18/U, CR-66/U, or similar.
  b. Load capacitance: 32 or 30 pF.
  c. Drive level: 0.3–1.0 mW, depending on the frequency and crystal resistance.
  d. Factors affecting frequency stability are as follows:
    1. Temperature coefficient of crystal: ±0.005 percent from $-55°C$ to $+105°C$ for CR-18/U, ±0.002 percent for CR-66/U.
    2. Aging rate of crystal: See section 5.7.
    3. Temperature stability of oscillator circuitry: 2.6 parts in $10^8/°C$ at 10 MHz and 2 parts in $10^8/°C$ at 20 MHz.
    4. Voltage coefficient of oscillator: 2 parts in 10^7 at 20 MHz and 4 parts in 10^8 at 10 MHz for a 10-percent supply voltage variation.
  e. Output (see Note 1): For $R_L = 100\,\Omega$, e_1 is approximately 0.05–0.10 V, depending on frequency and crystal resistance. Voltage e_2 is in the range from 1.5 to 2.0 V.
  f. Permissible load: $100 \leqslant R_L \leqslant \infty$.
  g. Input power: 15 mW.

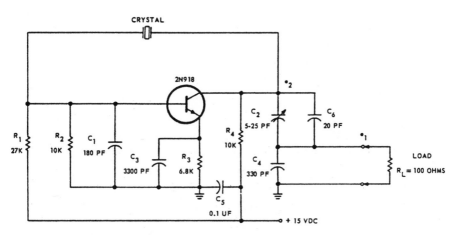

Figure 7-5. 10- to 30-MHz Pierce oscillator: schematic diagram.

NOTE 1. The output power can be increased by a factor of 3–5 by making several modifications. Change R_1 to 5.6 kΩ, R_3 to 10 kΩ, R_4 to 47 μH, C_1 to 220 pF, and R_L to 50 Ω. However, the frequency instability contribution of the oscillator circuitry is increased by roughly the same factor.

NOTE 2. Although the circuit will oscillate with any standard crystal in the 10- to 20-MHz range, some adjustment of C_2 is necessary over the frequency range to put crystals exactly on frequency.

7.2.6. Overtone Pierce Oscillator

The Pierce oscillator, as shown in Figure 7-2, cannot be used with overtone crystals. The circuit can oscillate with less gain on the fundamental frequency of the crystal than on an overtone. It is therefore necessary to provide some means of preventing the unwanted oscillation. If the collector capacitance is replaced by a tank circuit, this can be done conveniently. From equation (7-9) it can be seen that the reactances X_1 and X_2 must be of like sign for oscillation to take place. If the tank circuit is resonant between the fundamental and third overtone (or third and fifth overtones, if fifth-overtone crystals are to be used), oscillation of the unwanted modes will be prevented because the tank is inductive and has a positive sign. A capacitive reactance still appears at the desired frequency, however, and the desired oscillation can take place.

Equations (7-11) and (7-12) may be used to analyze the circuit provided the reactance of the tank circuit is substituted for X_2 in the equations. It can be shown that

$$X_2 = \frac{X_L}{1 - (f/f_r)^2},$$

where L and C_2 form the tank circuit, $X_L = 2\pi f L$, and $f_r = 1/2\pi \sqrt{LC_2}$. The effective output conductance of the transistor must be modified by adding the shunt conductance of the tank if it is not negligible. The tank conductance is given by $G_L = 1/QX_L$.

Above about 50 MHz, currently available transistors do not possess a sufficiently large y_{fe} to oscillate with 32-pF crystals and still maintain adequate reserve gain. If crystals are operated at a 20-pF load capacitance, the upper range can be extended to about 75 MHz. The reason for this can be seen from equations (7-11) and (7-12): $X_e \doteq -(X_1 + X_2)$, and since X_e for a given load capacitance is decreasing as the frequency increases, X_1 and X_2 also must be made to decrease. The product $X_1 X_2$ for a given value of $(X_1 + X_2)$ is largest when $X_1 = X_2$ and is given by $X_e^2/4$.

The transconductance required for oscillation then is given by:

$$g_{fe} = \frac{4R_e}{X_e^2}$$

This increases as the square of the frequency. Unfortunately, the effective resistance of a crystal increases as the load capacitance is made smaller according to the relationship

$$R_e = R_1 \left(\frac{C_L + C_0}{C_L}\right)^2$$

and a C_L of 20 pF is about the limit for crystals having a C_0 of 4 pF. Thus, the use of a parallel resonant Pierce oscillator is not practical above about 75 MHz. The use of any parallel resonant oscillator above 30 MHz has a serious disadvantage in that standard crystals which are ground to be on frequency at series resonance cannot be used. The parallel resonant Pierce oscillator has a major advantage, however, in that it is practically impossible for free-running oscillations to take place. Series resonant oscillators have a potential to free-run through the C_0 of the crystal if not designed carefully. Figure 7-6 is an example of a Pierce oscillator which can be used with overtone crystals.

7.2.7. 25-MHz Pierce Oscillator

Typical performance characteristics for the 25-MHz Pierce oscillator shown in Figure 7-6 are given below.

    a. Crystal: Fundamental or overtone.
    b. Maximum resistance: 50 Ω.
    c. Load capacitance: 32 pF.
    d. Drive level: 2 mW.

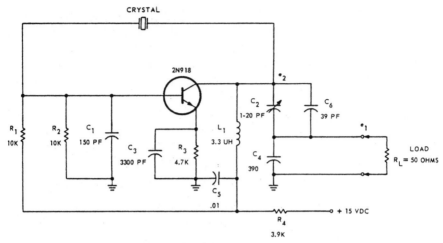

Figure 7-6. 25-MHz Pierce oscillator: schematic diagram.

e. Factors affecting frequency stability are as follows:
   1. Temperature coefficient of crystal: ±0.002 to ±0.005 percent from −55°C to +105°C, depending on crystal type.
   2. Aging rate of crystal: See section 5.7.
   3. Temperature stability of oscillator circuitry: 3 parts in 10^9/°C with third-overtone crystal.
   4. Voltage coefficient of oscillator: 2.3 parts in 10^7 for a 10-percent supply voltage change with overtone crystal.
f. Output: If R_L = 50 Ω, e_1 is approximately 0.1 V. The voltage e_2 is approximately 1.2 V. The output variation versus temperature is approximately ±6 percent from −55°C to +100°C.
g. Permissible load: $50 \leqslant R_L \leqslant \infty$.
h. Input power: 22 mW.

7.2.8. Impedance-Inverting Pierce Oscillator

The Pierce oscillator can be modified to use series resonant crystals. This is done by adding an inductor in series with the crystal to bring its frequency down to series resonance. Several advantages result from such a modification. First, standard, series resonant overtone crystals can be used. Also the reactances X_1 and X_2 can be made larger so that the transistor gain requirement is decreased. The power output also can be increased substantially.

The addition of a series inductor has several disadvantages, the most important of which is susceptibility to free running. If the inductor is relatively large, free-running oscillations may occur through the crystal C_0 instead of through the motional arm of the quartz. This is particularly true if the tank circuit is tuned so that the crystal frequency is being pulled high.

Free running can be alleviated if the crystal C_0 is resonated out by placing an inductor across it. It is then possible, however, for the oscillator to free-run below the crystal frequency through the C_0 compensating inductor. This sometimes occurs if the tank circuit is tuned so that the crystal frequency is being pulled low. Both of these free-running problems can be alleviated often by de-Q-ing the C_0 compensating inductor so that a circuit similar to that of Figure 7-8 (see section 7.2.10 below) results. Caution should be exercised here, since shunting the crystal with low impedances may increase the tendency of the oscillator to jump to a crystal spurious response.

Equations (7-11) and (7-12) may be used to analyze the series resonant Pierce oscillator if it is remembered that the quantity X_e includes both the reactance of the crystal and the inductor in series with it.

If a crystal switch with a significant amount of stray capacitance

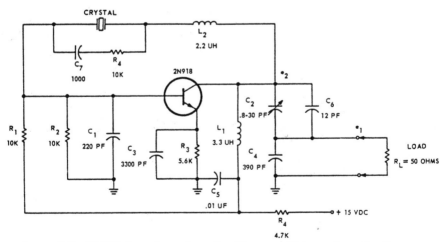

Figure 7-7. 25-MHz impedance-inverting Pierce oscillator: schematic diagram.

is to be used, it usually is better to locate the crystals and switch on the base side of the series inductor, as the circuit is less susceptible to stray capacitance on that side of the inductor.

In the design of VHF Pierce oscillators, excessive crystal drive is often a problem. This usually can be solved by decreasing the transistor emitter current until an acceptable drive level is obtained.

The Pierce oscillator family cannot be tuned for maximum output; they must be tuned for on-frequency operation. This may be a considerable disadvantage for field maintenance.

Figures 7-7, 7-8, and 7-9 are examples of impedance-inverting Pierce oscillators, and are included as a guide for designing such circuits.

7.2.9. 25-MHz Impedance-Inverting Pierce Oscillator

Typical performance characteristics for the 25-MHz impedance-inverting Pierce oscillator shown in Figure 7-7 are given below.

a. Crystal: Similar to CR-67/U.
b. Load capacitance: Series resonance.
c. Drive level: Nominally 2 mW.
d. Factors affecting frequency stability are as follows:
    1. Temperature coefficient of crystal: ±0.0025 percent.

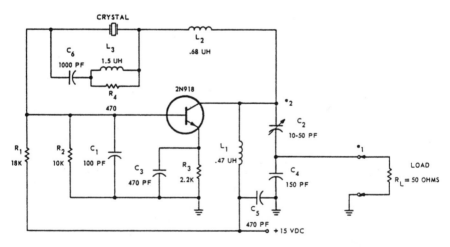

Figure 7-8. 50-MHz impedance-inverting Pierce oscillator: schematic diagram.

2. Aging rate of crystal: See section 5.7.
3. Temperature stability of oscillator circuitry: 1.6 parts in $10^8/°C$.
4. Voltage coefficient of oscillator: 1 part in 10^7 for a 10-percent variation in supply voltage.

e. Output: If R_L = 50 Ω, e_1 is approximately 0.15 V. Distortion is about 2.5 percent. The voltage e_2 is on the order of 2.5 V. Output voltage variation over the temperature range from $-55°C$ to $+105°C$ is approximately ±8 percent.
f. Permissible load: $50 \leqslant R_L < \infty$.
g. Input power: 20 mW.

7.2.10. 50-MHz Impedance-Inverting Pierce Oscillator

Typical performance characteristics for the 50-MHz impedance-inverting Pierce oscillator shown in Fig 7-8 are given below.

a. Crystal: Similar to CR-84/U.
b. Load capacitance: Series resonance.
c. Drive level: 1 mW.
d. Factors affecting frequency stability are as follows:
   1. Temperature coefficient of crystal: ±0.002 percent from $-55°C$ to $+105°C$.
   2. Aging rate of crystal: See section 5.7.
   3. Temperature stability of oscillator circuitry: 3 parts in $10^9/°C$.
   4. Voltage coefficient of oscillator: 1.5 parts in 10^7 for a 10-percent supply voltage variation.
e. Output: For R_L = 50 Ω, e_i is approximately 0.3 V, and e_2 is on the order of 1.5 V. The output voltage change with temperature is approximately ±10 percent from $-55°C$ to $+100°C$.
f. Permissible load: $50 \leqslant R_L \leqslant \infty$.
g. Input power: 40 mW.

7.2.11. 75-MHz Impedance-Inverting Pierce Oscillator

Typical performance characteristics for the 75-MHz impedance-inverting Pierce oscillator shown in Figure 7-9 are given below.

a. Crystal: Similar to CR-56/U.

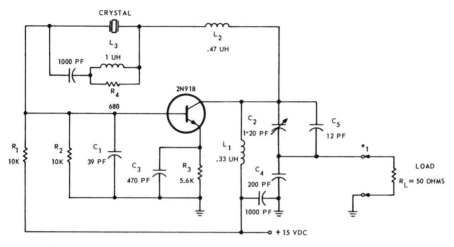

Figure 7-9. 75-MHz impedance-inverting Pierce oscillator: schematic diagram.

   b. Load capacitance: Series resonance.

   c. Drive level: Approximately 0.6 mW.

   d. Factors affecting frequency stability are as follows:

     1. Temperature coefficient of crystal: ±0.005 percent from −55°C to +105°C.

     2. Aging rate of crystal: See section 5.7.

     3. Temperature stability of oscillator circuitry: 4.2 parts in $10^9/°C$.

     4. Voltage coefficient of oscillator: 1 ppm for a 10-percent supply voltage variation.

   e. Output: For R_L = 50 Ω, e_1 is approximately 0.05 V. The output voltage change with temperature is approximately ±15 percent from −55°C to +100°C.

   f. Permissible load: $50 \leqslant R_L \leqslant \infty$.

   g. Input power: 30 mW.

7.3. COLPITTS OSCILLATOR

The Colpitts oscillator is actually a Pierce oscillator with the collector rather than the emitter at ac ground. If appropriate allowances are made for strays, the Pierce oscillator equations can be used for this circuit also. It may be desirable, however, to look at the Colpitts

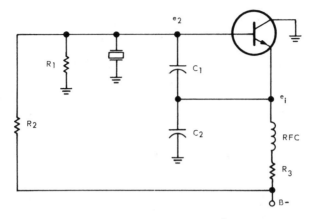

Figure 7-10. Colpitts crystal oscillator: schematic diagram.

circuit as a unique circuit analyzed in its own way. This gives a better feel for the quantities involved. (See Figure 7-10.) The Colpitts oscillator behaves quite differently from the Pierce oscillator in certain respects. The most important difference is in the biasing arrangement, which may present problems for the Colpitts circuit. The resistors R_1 and R_2 are not entirely negligible for low frequencies and may have several degrading effects. They may increase the effective resistance of the crystal branch of the circuit, thus reducing its Q in addition to decreasing the loop gain. They also may cause relaxation-type oscillations under certain conditions. Both of these problems can be reduced by using field effect transistors (FET). Temperature stability is somewhat worse with an FET, however.

The Colpitts oscillator can be thought of as an emitter-follower and a capacitive tapped tank circuit, as shown in Figure 7-11.

If capacitors C_1 and C_2 are large enough so that the input and output impedances of the transistor are effectively swamped, and if the crystal resistance R_e is small, then it can be shown (see Appendix F) that the step-up ratio of the tank circuit is

$$\frac{e_2}{e_1} \doteq \frac{X_1 + X_2}{X_2} \tag{7-28}$$

with voltages e_1 and e_2 in phase. (For this to be true, the crystal is

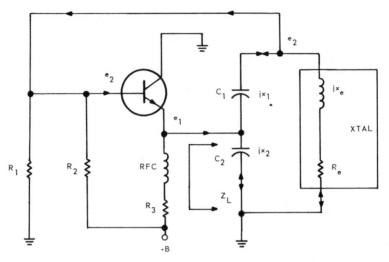

Figure 7-11. Colpitts oscillator: signal flow diagram.

resonant with the series combination of C_1 and C_2.) It also can be shown (see Appendix F) that the load seen by the emitter-follower is

$$Z_L \doteq \frac{X_2^2}{R_e} \tag{7-29}$$

and is purely resistive. The gain of the emitter-follower is

$$A = \left(\frac{e_1}{e_2} \right) = \frac{g_m Z_L}{1 + g_m Z_L}. \tag{7-30}$$

Since Z_L is resistive, the phase shift through the emitter-follower is zero, and the phase shift of the entire loop is zero. Oscillation then will take place if the loop gain exceeds unity. This is true if

$$\frac{e_2}{e_1} \, A \geqslant 1. \tag{7-31}$$

Substituting for A and e_2/e_1 gives

$$\left(\frac{g_m Z_L}{1 + g_m Z_L} \right) \left(\frac{X_1 + X_2}{X_2} \right) \geqslant 1 \tag{7-32}$$

Then substituting for Z_L gives

$$\left[\frac{g_m X_2^2/R_e}{1 + g_m X_2^2/R_e}\right]\left(\frac{X_1 + X_2}{X_2}\right) \geqslant 1. \qquad (7\text{-}33)$$

But simplifying this gives

$$g_m X_1 X_2 \geqslant R_e, \qquad (7\text{-}34)$$

which is the same equation obtained for the Pierce oscillator, as might be expected. Also, since the crystal must be resonant with the series combination of C_1 and C_2, the crystal reactance can be calculated using the equation $X_1 + X_2 + X_e = 0$, where X_e is the crystal reactance.

This analysis is quite limited by the assumptions made. When any one of them is not satisfied, equations (7-11) and (7-12) developed for the Pierce oscillator may be used. Here again the linear analysis gives no information concerning output voltage or crystal drive.

In some cases, the Y-parameters of the transistor may not be known or, for other reasons, the reader may elect to use an experimental approach to designing a Colpitts oscillator. For such an approach, the following guidelines may be used: In general, C_1 and C_2 should be as large as possible but still allow the circuit to oscillate with two to three times the maximum permissible crystal resistance.* If this results in the crystal reactance X_e being smaller than that of the specified load capacity, a trimmer capacitor may be placed in series with the crystal to trim the crystal onto frequency. As in the Pierce oscillator, it is desirable to let $|X_2| \ll 1/g_{oe}$ and $|X_1| \ll 1/g_{ie}$. This minimizes the effect of transistor input and output conductances on the circuit. Best stability also occurs if C_1 and C_2 are as large as possible because they swamp out any change in the transistor capacitances. Still another effect makes it desirable to make C_1 and C_2 as large as possible in the Colpitts oscillator. Referring to Figure 7-10, it can be seen that the reactance of the crystal (or the resultant reactance of the crystal and a series trimmer) will be smaller

---

*This can be determined by adding resistance in series with the crystal until oscillation will not occur.

if C_1 and C_2 are large, thus minimizing the shunt effects of R_1 and R_2.

The large-signal analysis presented for the Pierce oscillator in section 7.2.2 can also be applied to the Colpitts oscillator by recognizing that if the ac ground point were moved from the collector to the emitter, the circuits would be basically the same.

Again, as in the case of the linear analysis, additional insight may be obtained from an analysis based on the Colpitts configuration itself. Such an analysis has been made in Appendix I, based on the principle of harmonic balance. This analysis also shows an effect of amplitude on the frequency of oscillation. The analysis is made using the circuit of Figure 7-12. Here the crystal is replaced by an equivalent resistance and inductance. While this substitution would not be valid for a transient analysis or an analysis based on the variation of parameters, it is nevertheless satisfactory using the principle of harmonic balance, even though differential equations are used initially. The justification for this rests on the argument that the principle of harmonic balance is a steady-state solution, and replacing the crystal reactance by its series capacitance–inductance combination would only result in a slower buildup of oscillation. One might wonder if the resultant additional filtering afforded by the series LC combination would affect the result, since harmonics could beat together to produce a fundamental component. It should be observed, however, that even with the equivalent circuit shown, the impedance of the resonant circuit is so low at the harmonic frequencies that these components have a negligible effect on the amplitude of oscillation.

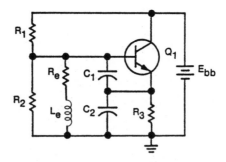

Figure 7-12. Colpitts oscillator circuit.

In the analysis, it is assumed that the biasing resistors R_1, R_2, and R_3 have a negligible effect on the ac performance. It will be found in general that the nonlinear analyses become formidable when even a few effects are considered, and therefore it is impossible to consider all the effects simultaneously. The general approach is to consider only one or two effects in a given analysis to determine how these parameters influence behavior. Other analyses are then made to determine how some of the neglected effects modify the behavior. For this analysis, it is assumed that the transistor reactances and its output impedance are negligible.

From equation (I-36) in Appendix I, we see that the amplitude of oscillation is determined by the expression:

$$X_1 X_2 g_m = R_e - \frac{X_1(X_e + X_2)}{R_{in}} \qquad (7\text{-}35)$$

where $X_1 = -1/\omega C_1$, $X_2 = -1/\omega C_2$, X_e is the crystal equivalent reactance, and g_m and R_{in} are the equivalent transconductance and input resistance at the final stabilized amplitude. The relationship between the small-signal values and the large-signal equivalent values is consistent with the nonlinear analysis discussed in section 6.4, and the ratios g_m/g_{m0} and R_{in}/R_{in0} may be read from the graph of Figure 6-6. As a first approximation, the last term of equation (7-35) may be neglected and the required value of g_m determined. Then from transistor data or by using the approximation $g_{m0} = qI_e/KT = 0.04I_e$, where I_e is in milliamperes, we find the ratio g_m/g_{m0}. Then using Figure 6-6, a value of V can be determined. This value can be used to read the ratio R_{in}/R_{in0}. The value of R_{in} can then be calculated.

The analysis of Appendix I, equation (I-28), also shows that the frequency of oscillation must satisfy the expression:

$$X_e + X_1 \left(1 + \frac{R_e}{R_{in}}\right) + X_2 = 0 \qquad (7\text{-}36)$$

where R_{in} is the equivalent input resistance at the final amplitude as defined above. From this equation, the crystal reactance X_e can be calculated (or if it is fixed, the values of X_1 and X_2 to obtain X_e can be found). The values of X_1, X_2, X_e, and R_{in} can then be substituted in equation (7-35) to obtain a corrected value of g_m if required.

The value of V can also be used to find the actual voltage across C_1, which is

$$V_1(\text{peak ac}) = \frac{VKT}{q} \doteq 26 \text{ mV times } V.$$

Since the circulating current in the tank circuit, consisting of R_e, L_e, C_1, and C_2, is normally large compared to the base current, the ac voltage across C_2 is given by

$$V_2 = \frac{V_1 C_1}{C_2}.$$

The bias shift due to oscillation from equation (6-23) is given by

$$V_{\text{bias}} = \frac{KT}{q} \ln I_0(V)$$

and may be read from the graph of Figure 6-7.

Several typical examples of Colpitts crystal oscillators are given in Figures 7-13 and 7-14. They are included as guidelines for designing circuits of this type. For these circuits, the output was taken from the emitter through a capacitive divider. It may be convenient to take the output from another point in the circuit if a larger voltage is required and a high-impedance load exists.

7.3.1. 3- to 10-MHz Colpitts Oscillator *

Typical performance characteristics for the 3- to 10-MHz Colpitts oscillator shown in Figure 7-13 are given below.

 a. Crystal: CR-18/U or similar.
 b. Load capacitance: 32 pF.
 c. Drive level: 2–10 mW, depending on the frequency and crystal resistance.
 d. Factors affecting frequency stability are as follows:
  1. Temperature coefficient of crystal: ±0.005 percent from −55°C to +105°C.
  2. Aging rate of crystal: See section 5.7.
  3. Temperature stability of oscillator circuitry: 4.4 parts in $10^8/°C$ at 10 MHz and 6.3 parts in $10^8/°C$ at 3 MHz.

*See footnote p. 66.

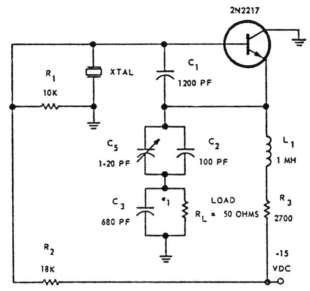

Figure 7-13. 3- to 10-MHz Colpitts oscillator: schematic diagram.

4. Voltage coefficient of oscillator: 1.3 parts in 10^7 at 3 MHz, 3 parts in 10^7 at 5 MHz, and 6 parts in 10^7 at 10 MHz for a 10-percent supply voltage variation.

e. Output: If $R_L = 50$ Ω, e_1 is approximately 0.05 V at 10 MHz and 0.25 V at 3 MHz, depending on the crystal resistance. Distortion is about 7 percent at 10 MHz and 17 percent at 3 MHz.

f. Permissible load: $50 \leqslant R_L \leqslant \infty$.

g. Power input: 55 mW.

NOTE. Although the circuit will oscillate with any standard crystal in the 3- to 10-MHz range, some adjustment of C_5 is necessary over the frequency range to put crystals exactly on frequency.

7.3.2. 10- to 20-MHz Colpitts Oscillator

Typical performance characteristics for the 10- to 20-MHz Colpitts oscillator shown in Figure 7-14 are given below.

a. Crystal: CR-18/U, CR-66/U, or similar.

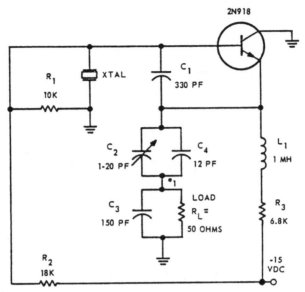

Figure 7-14. 10- to 20-MHz Colpitts oscillator: schematic diagram.

b. Load capacitance: 32 or 30 pF, respectively.
c. Drive level: 1–2 mW, depending on frequency and crystal resistance.
d. Factors affecting frequency stability are as follows:
   1. Temperature coefficient of crystal: ±0.005 percent from −55°C to +105°C for CR-18/U or ±0.002 percent from −55°C to +105°C for the CR-66/U.
   2. Aging rate of crystal: See section 5.7.
   3. Temperature stability of oscillator circuitry: 4.4 parts in $10^8/°C$ at 10 MHz and 1.8 parts in $10^8/°C$ at 20 MHz.
   4. Voltage coefficient of oscillator: Approximately 1.5 parts in 10^7 for a 10-percent change in supply voltage.
e. Output: For R_L = 50 Ω, e_1 = 0.15–0.20 V, depending on frequency and crystal resistance.
f. Permissible load: $50 \leqslant R_L \leqslant \infty$.
g. Input power: 26 mW.

NOTE. Although the circuit will oscillate with any standard crystal in the 10- to 20-MHz range, some adjustment of C_2 is necessary over the frequency range to put crystals exactly on frequency.

7.4. CLAPP OSCILLATOR

The Clapp oscillator is actually a Pierce oscillator with the base rather than the emitter at ac ground. If appropriate allowances are made for strays, then the Pierce oscillator equations can be used for this circuit. It may be desirable, however, to look at the Clapp oscillator as a unique circuit analyzed in its own way. This gives a better feel for the quantities involved. The most important disadvantage of the Clapp oscillator is that free-running oscillations may occur through the RF choke if resistor R_4 is too small. (See Figure 7-15.) If a fairly high supply voltage is available, R_4 can be made so large that the choke is not needed.

The Clapp oscillator can be thought of as a grounded-base amplifier stage loaded with a tank circuit. The tank has a capacitive tap from which energy is fed back to the emitter. Refer to Figure 7-16 for a signal diagram.

If we assume that the emitter base capacitance is included in C_1, that the collector-to-emitter capacitance is included in C_2, and that $R_e \ll X_e$, then the circuit can be analyzed as follows. It is assumed that the input impedance of the common-base amplifier is $1/g_m$, where g_m is the transconductance of the transistor. Also the gain of the stage

$$\frac{e_1}{e_2} = A \doteq g_m Z_L$$

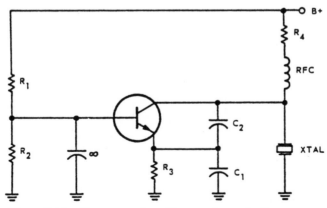

Figure 7-15. Clapp crystal oscillator: schematic diagram.

where Z_L is the collector load impedance. The voltage ratio of the tapped tank circuit is given approximately (see Appendix D) by

$$\frac{e_2}{e_1} = \frac{X_1}{X_1 + X_2}. \tag{7-38}$$

The phase angle between the voltages is very small; for practical purposes, the voltages e_1 and e_2 are in phase. The phase shift through the amplifier is determined by the phase angle of Z_L. It can be shown (see Appendix D) that Z_L, the impedance presented to the collector by the tank when the capacitive tap is loaded by the transistor input impedance $1/g_m$, is given by

$$Z_L \doteq \frac{(X_1 + X_2)^2}{R_e + g_m X_1^2} \tag{7-39}$$

This expression is derived under the condition that $X_1 + X_2 + X_e = 0$. Here, again, the phase angle is very small and, for this discussion, will be neglected. With a resistive load, then, the phase shift through the amplifier is zero; therefore, the phase shift through the entire loop is zero, fulfilling the phase shift requirements. The gain requirement is that the quantity $(e_2/e_1)A \geqslant 1$. The gain of the transistor is

$$A = g_m Z_L = \frac{g_m (X_1 + X_2)^2}{R_e + g_m X_1^2} \tag{7-40}$$

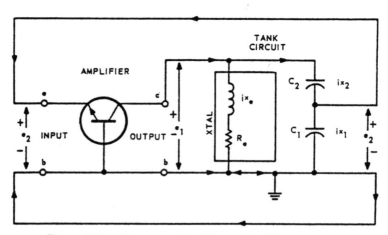

Figure 7-16. Clapp crystal oscillator: signal flow diagram.

and

$$\frac{e_2}{e_1} = \frac{X_1}{X_1 + X_2}.$$

Substituting these,

$$\left(\frac{X_1}{X_1 + X_2}\right)\left[\frac{g_m(X_1 + X_2)^2}{R_e + g_m X_1^2}\right] \geqslant 1. \qquad (7\text{-}41)$$

This fortunately simplifies to the equation derived for the Pierce oscillator: $g_m X_1 X_2 \geqslant R_e$. Also, it has been shown that the phase shift requirements will be fulfilled if $X_1 + X_2 + X_e = 0$.

This analysis is quite limited by the assumptions made. When any one of them is not satisfied, equations (7-11) and (7-12) developed for the Pierce oscillator may be used.

Here again the linear analysis gives no information concerning output voltage or crystal drive. Also, in some cases the Y-parameters of the transistor may not be known, and experimental design techniques must be used.

In general, C_1 and C_2 should be as large as possible but still allow the circuit to oscillate with two to three times the maximum permissible crystal resistance.* If this results in the crystal reactance X_e being smaller than that required for the specified load capacitance, a trimmer capacitor may be placed in series with the crystal. Whether or not a series capacitor is used, it may be desirable to provide some variable reactive element to trim the crystal onto frequency.

It should be noted that when C_1 and C_2 are large, the sum $X_1 + X_2$ is small, making the RF choke in series with R_4 unnecessary under some conditions.

A load can be connected to the Clapp oscillator in any one of several places. Maximum voltage can be obtained on the collector. Any impedance connected to this point must be high. A lower impedance and moderate voltages can be found at the emitter. A very low impedance output can be obtained by inserting a capacitor equal to approximately five times the value of C_1 between C_1 and

---

*This can be determined by adding resistance in series with the crystal until oscillation will not occur.

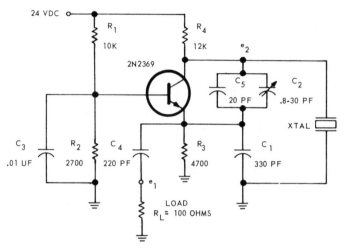

Figure 7-17. 3- to 20-MHz Clapp oscillator: schematic diagram.

ground. The output then is taken from the connection of C_1 and the added capacitor.

The schematic diagram of a practical Clapp oscillator is given in Figure 7-17.*

7.4.1. 3- to 20-MHz Clapp Oscillator Circuit

a. Crystal: CR-18/U, CR-66/U, or similar.
b. Load capacitance: 32 or 30 pF, respectively.
c. Crystal drive level: 1–5 mW, depending on frequency and crystal resistance.
d. Factors affecting frequency stability are as follows:
  1. Temperature coefficient of crystal: ±0.005 percent from −55°C to +105°C for CR-18/U or ±0.002 percent from −55°C to +105°C for CR-66/U.
  2. Aging rate of crystal: 5 parts in 10^6/month to several parts in 10^8/month, depending on crystal. See section 5.7.
  3. Temperature coefficient of oscillator: 3 parts in 10^8/°C at 3 MHz, 7 parts in 10^9 at 5 MHz, and 2.4 parts in 10^8 at 20 MHz.
  4. Voltage coefficient: 2 parts in 10^6 at 3 MHz, 9 parts in 10^7

*See footnote p. 66.

at 5 MHz, 4 parts in 10^7 at 10 MHz, and 6 parts in 10^7 at 20 MHz for a 10-percent change in supply voltage.

e. Output conditions: With R_L = 100 Ω, $e_1 \doteq 0.15$–0.3 V, depending on frequency and crystal resistance. Distortion (at 5 MHz) = 15 percent. The output may be taken from the collector with a very high-impedance load. Voltage $e_2 \doteq 3$–7 V, depending on frequency and crystal resistance. The output change over the temperature range from $-55°$C to $+100°$C is approximately 5 percent.

f. Permissible load: $100 \leqslant R_L < \infty$.

g. Input power: 65 mW.

Note 1. The oscillator can be operated from a 15-V supply if several changes are made. The resistor R_4 is decreased to 5.6 kΩ, and a 500-μH choke is added in series with it. The choke is necessary to prevent unduly loading the collector circuit. The resistor in series with the choke is necessary to prevent the oscillator from free-running through the choke instead of oscillating through the crystal. The emitter resistor R_e is reduced to 2.7 kΩ.

Note 2. Although the circuit will oscillate with any standard crystal in the 3- to 20-MHz range, some adjustment of C_2 is necessary over the frequency range to put crystals exactly on frequency.

7.5. GROUNDED-BASE OSCILLATOR

The basic circuit of the grounded-base crystal oscillator is shown in Figure 7-18.

This circuit can be used from several megahertz to above 150 MHz. It is most commonly used in the range from 20 to 100 MHz. The circuit is capable of delivering high output power, has medium frequency stability, and is about average in difficulty to design. It is basically a zero phase shift oscillator. This makes it undesirable for use with a complicated crystal switch in the region above 75 MHz where lagging phase shift problems become severe.

Basically, the grounded-base oscillator circuit works as follows: The voltage on the emitter of the transistor is amplified and appears

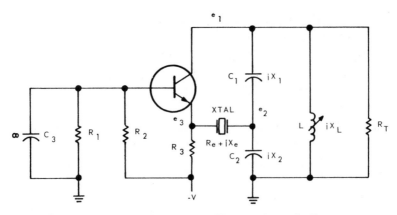

Figure 7-18. Grounded-base oscillator: schematic diagram.

on the collector as

$$e_1 = e_3 y_{fb} Z_t \qquad (7\text{-}42)$$

where Z_t is the impedance presented to the collector by the tank. For most oscillators, the tank is tuned to resonance and thus appears resistive. The voltage e_1 then lags e_3 by the angle of the forward transfer admittance of the transistor, which at lower frequencies is small.

A small amount of phase shift also is caused by the capacitive tap on the tank circuit. This is a leading phase shift, however, and tends to compensate for the phase lag in the transistor. If the capacitor C_2 is fairly large, this phase shift is negligible and the voltage on the tap is given approximately by

$$e_2 = e_1 \left(\frac{C_1}{C_1 + C_2} \right) = -e_1 \left(\frac{X_2}{X_L} \right). \qquad (7\text{-}43)$$

The grounded-base input impedance of the transistor normally is somewhat inductive so that leading phase shift also occurs between the tap on the tank e_2 and the emitter. The emitter voltage e_3 is related to e_2 by the expression

$$e_3 = \left(\frac{e_2}{1 + Z_e y_{ib}} \right) \qquad (7\text{-}44)$$

where Z_e is the crystal impedance and y_{ib} is the transistor input admittance. If the crystal is at series resonance, and if the transistor input admittance is only slightly inductive, then the expression simplifies to

$$e_3 \doteq e_2 \left(\frac{R_{in}}{R_{in} + R_e} \right) \qquad (7\text{-}45)$$

where R_{in} is the input resistance of the transistor and R_e is the crystal resistance. The phase shift that does occur tends to compensate for the phase lag in the transistor. For oscillation to take place, all three of the phase shifts must add up to zero, and the loop gain must equal or exceed unity.

The equations which must be satisfied for these conditions to occur are derived in Appendix C and are presented here as equations (7-46) and (7-47).

$$g_{fb} = \frac{1}{R_T} \left(\frac{X_L}{X_2} \right) \left[\frac{R_e + R_{in}}{R_{in}} \right] + \frac{1}{R_{in}} \left(\frac{X_2}{X_L} \right)$$

$$+ b_{ib} \left(\frac{X_1}{R_T} \right) - b_{ib} \left(\frac{X_L}{X_2} \right) \left(\frac{X_e}{R_T} \right) \quad (7\text{-}46)$$

$$X_e = b_{fb} R_T R_{in} \left(\frac{X_2}{X_L} \right) + X_1 \left(\frac{X_2}{X_L} \right) - b_{ib} R_{in} \left[R_e + R_T \left(\frac{X_2}{X_L} \right)^2 \right]$$

$$(7\text{-}47)$$

where $X_1 = -1/\omega C_1$, $X_2 = -1/\omega C_2$, $X_L = \omega L$, $R_{in} = 1/g_{ib}$, and R_T is the total resistance shunting the tank circuit ($R_T \doteq R_L$). (For definitions of the g and b transistor parameters, refer to Chapter 6.

In deriving these equations, the following assumptions were made:

a. The reverse transfer admittance of the transistor, y_{rb}, is negligible.

b. The transistor output admittance, y_{ob}, is either negligible or lumped in with R_T and L.

c. The tank components $X_1 + X_2 + X_L = 0$. (This is approximately resonance.)

Even with these assumptions, the equations appear formidable but are nevertheless quite useful. If equation (7-46) is minimized for transistor gain, it will be found that minimum g_{fb} is required when

$$\frac{X_L}{X_2} = -\left(\frac{R_T}{R_e + R_{in}}\right)^{1/2}. \tag{7-48}$$

This assumes that the crystal is operated at series resonance, $X_e = 0$. The minimum g_{fb} is then given by

$$g_{fb(min)} = -\frac{2}{R_{in}}\left(\frac{R_e + R_{in}}{R_T}\right)^{1/2} + b_{ib}\left(\frac{X_1}{R_T}\right). \tag{7-49}$$

For a first approximation, it may be desirable to neglect the reactive component b_{ib} of the transistor input impedance in the gain equation so that

$$g_{fb(min)} \doteq -\frac{2}{R_{in}}\left(\frac{R_e + R_{in}}{R_T}\right)^{1/2}. \tag{7-50}$$

If

$$\left(\frac{b_{ib}}{g_{fb}}\right)\left(\frac{X_1}{R_T}\right) \ll 1 \tag{7-51}$$

then the error introduced by this assumption is negligible, as is often the case.

If we require the crystal to operate at series resonance ($X_e = 0$), then we can design the oscillator circuit as follows by the manipulation of equations (7-47), (7-48), and (7-50).

Step 1. Calculate R_T using g_{fb} approximately one-third the actual g_{fb} of the transistor. This gives a loop gain of 3 to ensure saturation:

$$R_T = \frac{4(R_e + R_{in})}{(R_{in} \, g_{fb})^2}. \tag{7-52}$$

Step 2. Calculate (X_L/X_2) using the equation,

$$\frac{X_L}{X_2} = -\left(\frac{R_T}{R_e + R_{in}}\right)^{1/2}. \tag{7-53}$$

Note that

$$\frac{C_2}{C_1} = -\left(\frac{X_L}{X_2} + 1\right).$$

Step 3. Calculate X_1 (using b_{fb} approximately one-third the actual b_{fb} if one-third the actual g_{fb} was used in step 1):

$$X_1 = \frac{b_{ib}}{g_{ib}} \left[R_e \left(\frac{X_L}{X_2}\right) + R_T \left(\frac{X_2}{X_L}\right) \right] - \frac{b_{fb}}{g_{ib}} R_T. \qquad (7\text{-}54)$$

This equation must be used with judgment. If $g_{fb} < b_{fb}$ and R_T is fairly large, the calculated value of X_1 will be very large (small C_1). This results because of the assumption $X_1 + X_2 + X_L = 0$. If the transistor phase lag is large, it may require that a considerable phase shift be obtained in C_2. This causes the tank to look capacitive if $X_1 + X_2 + X_L = 0$, producing even more lagging phase shift and requiring C_1 to be extremely small. It may be better to keep C_1 a little larger and tune the tank to actual resonance (so that Z_T is real).

At low frequencies, the calculated value of X_1 may come out very small (C_1 extremely large) or negative. This results it the phase shift at the transistor input is sufficient to cancel the phase lag through the transistor so that no phase shift is required from the capacitive tap. If this occurs, it may be better to let

$$X_1 = \frac{R_e + R_{in}}{\text{approximately 5 or 10}} \left[\frac{X_L}{X_2} + 1 \right] \qquad (7\text{-}55)$$

and tune the tank slightly capacitive.

Step 4. Calculate X_2 by

$$X_2 = - \frac{X_1}{(X_L/X_2) + 1}. \qquad (7\text{-}56)$$

Step 5. Calculate X_L by

$$X_L = -(X_1 + X_2). \qquad (7\text{-}57)$$

If the parameters of the transistor are accurately known, then the application of equations (7-52) to (7-57) may lead to values for C_1, C_2, L, and R_T close to the final values in the optimized circuit. Since the equations assume linearity, they give no information concerning crystal drive level or output power. They predict starting conditions only. The power gain approach described in Chapter 3 may be useful when designing a grounded-base oscillator. If it is used, the transformation ratio $R_{fb}/(R_{in} + R_e)$, should be set equal to the transformation ratio of the tank circuit which is given by $[(C_1 + C_2)/C_1]^2$. This can be solved for the ratio

$$\frac{C_2}{C_1} = \left(\frac{R_{fb}}{R_{in} + R_e}\right)^{1/2} - 1. \qquad (7\text{-}58)$$

Regardless of which design approach is used, it will be necessary in general to optimize the circuit experimentally. Therefore, a general discussion for the experimental approach is given here.

Basically, there are three considerations which must be kept in mind in designing a grounded-base oscillator. First, the impedance transformation $[(C_2/C_1) + 1]^2$, should be approximately equal to the ratio $R_T/(R_{in} + R_e)$ for optimum gain conditions. R_{in} is the input resistance of the transistor and is generally quite low (in the range from 20 to 100 Ω). R_e is the crystal resistance and in the VHF range usually falls between 20 and 60 Ω. R_T is the collector load resulting from the load resistor and the tank circuit. If high output is required, it is desirable to make the ratio C_2/C_1 fairly large. This reduces the crystal dissipation for a given output voltage. It also reduces the stability; therefore, if the additional output power is not required and crystal dissipation is excessive, the emitter current should be reduced. If the emitter current is high and the ratio of C_2/C_1 is very large, outputs in excess of 50 mW can be obtained. However, outputs below 5 mW are more common for stable oscillators.

The second consideration in designing a grounded-base oscillator is the adjustment of C_1 and C_2. They should be adjusted so that the crystal is on frequency when the tank is tuned for maximum output (resonance). If there is too much phase lag in the transistor, the crystal will operate below series resonance. This usually can be corrected by decreasing C_1 and C_2 (the ratio C_2/C_1 may remain unchanged). The amount of phase shift that can be compensated for in this manner is somewhat limited. For this reason, the grounded-base oscillator is not desirable for use with complicated crystal switches above about 75 MHz. On occasions it may be desirable to insert a capacitor in series with the crystal to get the frequency up to series resonance.

A third consideration in designing a grounded-base oscillator is preventing unwanted oscillations. They may occur simultaneously with the crystal oscillation or they may be sufficiently severe to kill the crystal-controlled oscillation altogether. There are generally two types of free-running oscillation which may occur in the grounded-base oscillator. The first is oscillation through the shunt capacitance C_0 of the crystal rather than through the motional arm of the crystal. This usually can be prevented by resonating out C_0 by the addition of an inductor across the crystal, as shown in Figure 7-20 (see section 7.5.2 below). A second source of instability is the internal feedback of the transistor, which may cause parasitic oscillations. (Refer to Chapter 6.) These oscillations usually can be detected by a jump in the output voltage as the oscillator is tuned. The best remedy for such oscillations is to use a fairly small resistance value for R_T. Only if the actual load is resistive over a wide frequency range can resistor R_T be eliminated. The importance of using some real resistance to load the tank for stabilization cannot be overemphasized. In some cases, it may be desirable to load the emitter for stabilization also, as is done on the oscillator of Figure 7-19.

Some transistors have a considerably greater tendency than others to develop parasitic oscillations. Therefore, if the problem persists, it may be advantageous to try several other transistor types.

The grounded-base oscillator is sometimes used with an inductive tap rather than a capacitive tap for low-frequency crystals. This may make it easier to get the crystal down to its series resonant frequency. Adding a capacitor from emitter to ground also may be helpful in accomplishing this. Generally, the inductive tap should not be used above 30 MHz because it aggravates the lagging phase shift problem.

Several practical grounded-base oscillator circuits are presented in Figures 7-19 through 7-22.*

7.5.1. 25-MHz Grounded-Base Oscillator

$L = 0.65$–$1.5\ \mu H$, 14 turns of #28 wire close-wound on 0.211-inch-o.d. coil form. Slug: Carbonyl W, $\frac{3}{8}$ inch long.

C_1 has a negative temperature coefficient of 200 ppm/°C to compensate for temperature changes in the oscillator circuitry.

  a. Crystal: Similar to CR-67/U.
  b. Load capacitance: Series resonance.
  c. Drive level: Nominally 2 mW.
  d. Factors affecting frequency stability are as follows:
     1. Temperature coefficient of crystal: ±0.0025 percent from −55°C to +105°C.

*See footnote p. 66.

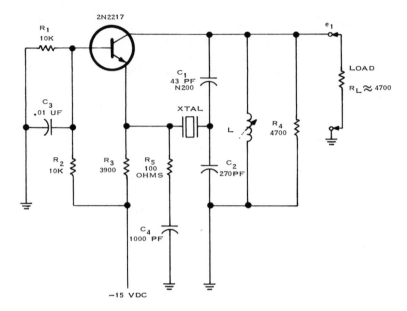

Figure 7-19. 25-MHz grounded-base oscillator: schematic diagram.

2. Aging rate of crystal: ±0.0005 percent/month to several parts in 10^8/month, depending on crystal. See section 5.7.
3. Temperature coefficient of oscillator circuitry: 1 part in 10^8/°C.
4. Voltage coefficient of oscillator: 7 parts in 10^7 for a 10-percent change in supply voltage.

e. Permissible load: 4.7 k$\Omega \leqslant R_L < \infty$.

f. Output: e_1 is approximately 2 V for R_L = 4.7 kΩ. Change in output with temperature is approximately 3 percent from -55°C to +100°C. A low impedance output may be obtained by using a capacitive divider in place of C_2.

NOTE. R_4 is used for stabilization and should not be included in the load.

g. Power input: 40 mW.

7.5.2. 50-MHz Grounded-Base Oscillator

L_1 = 0.92–2.1 μH, 15 turns #28 wire close-wound on 0.211-inch-o.d. coil form. Slug: Carbonyl W, $\frac{3}{8}$ inch long.

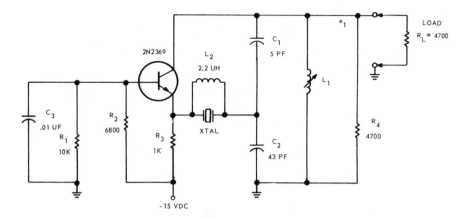

Figure 7-20. 50-MHz grounded-base oscillator: schematic diagram.

a. Crystal: Similar to CR-67/U.
b. Load capacitance: Series resonance.
c. Drive level: Nominally 2 mW.
d. Factors affecting frequency stability are as follows:
   1. Temperature coefficient of crystal: ±0.0025 percent from $-55°C$ to $+105°C$.
   2. Aging rate of crystal: See section 5.7.
   3. Temperature coefficient of oscillator circuitry: 1 part in $10^8/°C$.
   4. Voltage coefficient of oscillator: 3 parts in 10^7 for a 10-percent change in supply voltage.
e. Permissible load: $4.7 \text{ k}\Omega \leqslant R_L < \infty$.
f. Output: e_1 is approximately 7 V for a load of 4.7 kΩ. Change in output with temperature is approximately 5 percent from $-55°C$ to $+100°C$. A low impedance output may be obtained by using a capacitive divider in place of C_2. Distortion is approximately 5 percent.

NOTE. R_4 is used for stabilization and should not be included in the load.

g. Power input: 85 mW.

NOTE. L_2 is chosen to be antiresonant with the C_0 of the crystal and any stray capacitance in parallel with it.

7.5.3. 75-MHz Grounded-Base Oscillator

L_1 = 0.27–0.52 μH, 7 turns #28 wire close-wound on 0.211-inch-o.d. coil form. Slug: Carbonyl E, $\frac{5}{16}$ inch long.

L_2 = This inductance is chosen to be antiresonant with the C_0 of the crystal and any stray capacity in parallel with it.

  a. Crystal: Similar to CR-56A/U.

  b. Load capacitance: Series resonance.

  c. Drive level: Approximately 0.5 mW.

  d. Factors affecting frequency stability are as follows:

    1. Temperature coefficient of crystal: ±0.005 percent from $-55°C$ to $+105°C$.

    2. Aging rate of crystal: See section 5.7.

    3. Temperature coefficient of oscillator circuitry: 1 part in $10^8/°C$.

    4. Voltage coefficient of oscillator: 8 parts in 10^7 for a 10-percent change in supply voltage.

  e. Permissible load: 3.3 k$\Omega \leqslant R_L \leqslant \infty$.

  f. Output: e_1 is approximately 2 volts for a load of 3.3 kΩ. Change in output with temperature is approximately 5 percent from -55 to $+100°C$. A low impedance output may be obtained by using a capacitive divider in place of C_2. Distortion is approximately 10 percent.

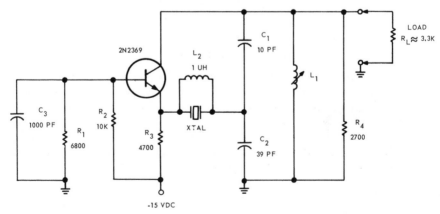

Figure 7-21. 75-MHz grounded-base oscillator: schematic diagram.

NOTE. R_4 is used for stabilization and should not be included in the load.

g. Power input: 40 mW.

7.5.4. 110-MHz Grounded-Base Oscillator and Doubler

L_1 = 0.20–0.40 μH, 7 turns #26 wire close-wound on 0.162-inch-o.d. coil form. Slug: Carbonyl SF, $\frac{3}{8}$ inch long.

L_2 is chosen to be antiresonant with the C_0 of the crystal and any stray capacitance in parallel with it.

L_3, C_4 : Omit if transistors with a higher f_t are used.

L_4: 2 $\frac{1}{2}$ turns #18 wire, $\frac{3}{16}$-inch-i.d., $\frac{9}{64}$ inch long.

C_1 has a negative temperature coefficient of 470 ppm/°C to compensate for temperature changes in the inductor, L_1 and in the remaining oscillator circuitry.

a. Crystal: Similar to CR-56A/U.
b. Load capacitance: Series resonance.
c. Drive level: Nominally 2 mW.
d. Spurious responses: 3 : 1 or 120 Ω, whichever is greater.
e. Pin-to-pin capacitance: 4.5 pF maximum.
f. Factors affecting frequency stability are as follows:
1. Temperature coefficient of crystal: ±0.005 percent from −55°C to +105°C.

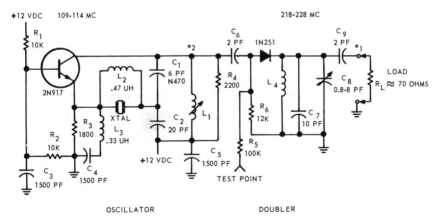

Figure 7-22. 110-MHz grounded-base oscillator and doubler: schematic diagram.

2. Aging rate of crystal: See section 5.7.
3. Temperature coefficient of oscillator circuitry: 2 parts in $10^8/°C$.
4. Voltage coefficient of oscillator: 8 parts in 10^7 for a 10-percent supply in voltage change.

g. Output: For $R_L = 70 \ \Omega$, e_1 is approximately 0.15–0.25 V at 220 MHz.

7.6. GATE OSCILLATORS

The use of logic gates in crystal oscillators is common in systems where the oscillator output must drive digital hardware. These oscillators are generally of lesser stability than those discussed previously in this chapter; however, they are very useful and a large variety of such oscillators have been used. Nearly all types of gate oscillators are prone to problems with respect to free running and spurious oscillations, and it is recommended that the oscillators be thoroughly tested in accordance with the procedures outlined in the following chapter prior to committing them to production. Several types of gate oscillators which have been found to work satisfactorily in some applications are discussed in the following paragraphs.

7.6.1. Single-Gate Oscillators

The single-gate oscillators, particularly in the lower frequency range using CMOS gates, have proven to be satisfactory in many applications. The frequency stability is generally not as good as that obtained with the transistor oscillators discussed earlier such as the Pierce and Colpitts circuits. As a result, gate oscillators are usually not used in temperature-compensated applications or oven frequency standards. When a frequency stability degradation of several parts per million from that of the crystal can be tolerated, the use of a single-gate oscillator is often a good choice.

A basic low-frequency gate oscillator circuit is shown in Figure 7-23 and consists of the gate $U1$ followed by a resistance R_2 to raise the effective output impedance of the gate. This combination of the gate and resistor may be thought of as replacing the transistor in a Pierce oscillator. Refer to Figure 7-1.

The gate provides the necessary gain and produces a phase shift of 180 degrees. The π-network, consisting of C_1, C_2, and the crystal, produces an additional 180 degrees of phase shift, thus satisfying the 360-degree phase shift requirements for oscillation.

The crystal looks inductive and is resonant with capacitors C_1 and C_2. The frequency of oscillation automatically adjusts itself so that this is true; therefore, the combination of the crystal and C_1 alone has a net inductive reactance at the operating frequency.

Current I_1 lags voltage e_2 by 90 degrees. Voltage e_1 being developed across capacitor C_1 lags current I_1 by 90 degrees, making it 180 degrees behind voltage e_2. Since the combination of C_1 and the crystal is resonant with C_2, the gate, through R_2, sees a resistive load.

This explanation is valid only at low frequencies where the gate produces no phase shift, and if the input impedance of the gate is negligible compared to the output impedance of the π network. In the more usual case some phase compensation for these effects is necessary and occurs at the input of the π network due to the presence of R_2 which then looks into a somewhat reactive load. The capacitors C_1 and C_2 are then not exactly resonant with the crystal. In some gate oscillators the input impedance of the gate significantly loads the π network which also reduces the maximum resistance that the circuit can accommodate. At low frequencies, the single gate oscillator can usually be designed to accept the maximum crystal resistance with no difficulty. A more difficult aspect of the design seems to be the elimination of spurious and relaxation type oscilla-

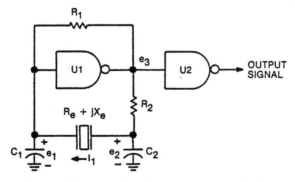

Figure 7-23. Basic low-frequency gate oscillator.

tion. In some cases during the design the circuit may actually fail to oscillate on the crystal frequency but produce relaxation type oscillation at some other frequency. In other cases the circuit will oscillate on the crystal frequency with an envelope at another frequency appearing on the signal. In other cases, and this is perhaps the most evasive, the circuit will operate satisfactorily with a typical crystal but will cause spurious oscillations with a high resistance crystal or if the crystal is removed. These conditions should be evaluated prior to completing the design and if not permissible in a particular application they must be eliminated.

It is often rather difficult to completely eliminate the tendency for these free running oscillations and a great deal of component value modification may be necessary, as well as layout changes and improvements in the supply voltage bypassing. It has been found that in some cases that the use of a second gate for load isolation, as shown in Figure 7-23 may result in relaxation oscillation tendencies and the choice of which gate on the chip is used may have an impact on the performance.

Even though the gate oscillators are plagued with these spurious tendencies they are widely used and have proven to be satisfactory in many applications. The designer must be aware of the potential problems and even though the circuits appear simple and straight forward to design, the design effort should not be terminated when the circuit first starts to oscillate.

The circuit can be analyzed analytically by substituting the Y-parameters of the gate with the resistor R_2 on its output into the equations for the Pierce oscillator given in section 7.2.

In section 7.2 it is shown that for oscillation to take place,

$$g_{fe} X_1 X_2 \geqslant R_e + K_1 \qquad (7\text{-}59)$$

and

$$X_1 + X_2 + X_e = 0 + K_2, \qquad (7\text{-}60)$$

where K_1 and K_2 are second-order corrections given by:

$$K_1 = -X_1(X_2 + X_e)g_{ie} - X_2(X_1 + X_e)g_{oe}$$
$$-R_e X_1 X_2 (g_{ie}g_{oe} + b_{fe}b_{re}) - b_{re}g_{fe} X_1 X_2 X_e \quad (7\text{-}61)$$

and

$$K_2 = b_{fe}X_1X_2 + X_1X_2X_eg_{ie}g_{oe} - R_e(X_1g_{ie} + X_2g_{oe})$$
$$+ b_{re}X_1X_2 + b_{re}b_{fe}X_1X_2X_e - b_{re}g_{fe}X_1X_2R_e. \quad (7\text{-}62)$$

Assuming that the Y-parameters of the gate are known, we can determine the Y-parameters of the combination of the gate and a series output impedance Z. (Refer to Figure 7-24.)

We find that

$$y_i = y_{11} - \frac{y_{12}y_{21}Z}{1 + y_{22}Z}, \quad (7\text{-}63)$$

$$y_r = y_{12} - \frac{y_{12}y_{22}Z}{1 + y_{22}Z}, \quad (7\text{-}64)$$

$$y_f = \frac{y_{21}}{1 + y_{22}Z}, \quad (7\text{-}65)$$

and

$$y_o = \frac{y_{22}}{1 + y_{22}Z}. \quad (7\text{-}66)$$

If the reverse transfer admittance is negligible and $y_{22}Z = y_{22}R_2 \gg 1$, then equations (7-63) through (7-66) simplify to

$$y_i = y_{11} \quad (7\text{-}67)$$

$$y_r = 0 \quad (7\text{-}68)$$

$$y_f = \frac{y_{21}}{y_{22}R_2} \quad (7\text{-}69)$$

$$y_o = \frac{1}{R_2} \quad (7\text{-}70)$$

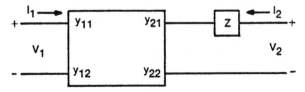

Figure 7-24. Gate with external output impedance.

We note also that the open-loop gain of a two-port is given by

$$A = -\frac{y_f}{y_o} \doteq -\frac{y_{21}}{y_{22}}; \qquad (7\text{-}71)$$

hence $y_f \doteq -A/R_2$. The gain of a typical CMOS gate is on the order of 20. The values used for R_2 may vary from perhaps 300 kΩ at 30 kHz to 10 kΩ at 300 kHz. The value of R_1 (Figure 7-23) is usually chosen in the 10- to 20-MΩ range to bias the gate in its active region. Circuit values for a 200 kHz oscillator constructed by the author are listed in Table 7-2.

Table 7-2. Circuit Values for Low-Frequency Gate Oscillator.*

Frequency (kHz)	Gate type	R_1 (MΩ)	R_2 (kΩ)	C_1 (pF)	C_2 (pF)
200.0	4049	22	12	120	24

*See footnote p. 66.

The 200-kHz circuit uses a 20-pF crystal. The voltage coefficient for this circuit was in the order of 3 pp 10^7 per percent of supply voltage change. The gate pulls the frequency approximately 7 pp $10^8/°C$. Power supply current including the buffer is 1.7 mA from a 5-V supply.

At higher frequencies, above approximately 1 MHz, the circuit of Figure 7-23 is not entirely satisfactory because of lagging phase shift in the gate, and it is necessary to replace resistor R_2 with a capacitor C_3, as shown in Figure 7-25. Equations (7-59) and (7-60) may still be used to analyze the circuit; however, equations (7-63) through (7-66) should be used to determine the Y-parameters of the gate with capacitor C_3 on the output. Z in this case is $-j/\omega C_3$. It should be noted that in the derivation of the equations for the Pierce oscillator, it was assumed that the input and output susceptances of the transistor were lumped into C_1 and C_2 so that b_{ie} and b_{oe} do not appear in the equations for oscillation. Here also b_{ie}/ω should be added to C_1; b_{oe}/ω to C_2.

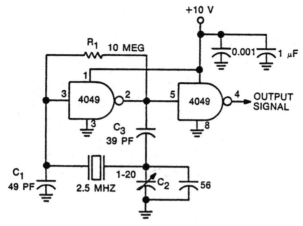

Figure 7-25. 2.5-MHz gate oscillator.

The circuit shown in Figure 7-25 uses a 32-pF parallel resonant crystal.

The voltage coefficient of the oscillator is on the order of 6 parts in 10^8/percent of supply voltage change. The voltage coefficient is somewhat dependent on the value of C_3 which should not be made smaller than perhaps 20 pF. The current drawn by the circuit is 6.5 mA from a 10-V supply or 1.5 mA from 6 V dc. The voltage at the gate output is 10 V peak to peak for 10 V supply and 6–10 V peak to peak at the gate input, depending on the crystal resistance. The gate pulls the frequency approximately 5 parts in 10^8/°C.*

For frequencies higher than a few MHz, it is necessary to use TTL gates for satisfactory operation. Unfortunately the large resistor R_1 is not adequate to bias a TTL gate into the active region. Low values of R_1 produce a considerable signal feedback and reduce the gain to an unacceptable level. The arrangement shown in Figure 7-26 is reasonably acceptable, however, there may be some tendency for free running and relaxation oscillation, which is a function of the gate and the circuit layout. The designer should particularly watch for these if he elects to use this type of circuit.

For the circuit of Figure 7-26, the gate pulls the frequency of the crystal less than 1 ppm over the temperature range from −40°C–

*See footnote p. 66.

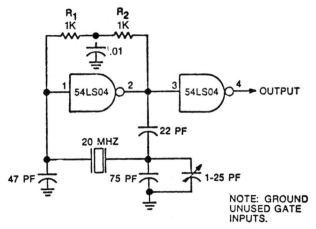

Figure 7-26. 20-MHz gate oscillator.

+65°C. This should be added to the basic crystal frequency tolerance, which is usually 25–50 ppm, depending on the crystal used.

The voltage coefficient was found to be 2–10 parts in 10^7 for a 20-percent supply voltage change.*

If the circuit is used at lower frequencies, C_1, C_2, and C_3 should be increased to ensure operation on the fundamental mode of the crystal. A series capacitor can then be used with the crystal to tune it to frequency. Since this circuit is quite prone to free-running oscillation, particularly with small values of C_1, it should be thoroughly tested with maximum-resistance crystals and limit-of-tolerance parts.

7.6.2. Multiple-Gate Oscillators

Multiple-gate oscillators, usually using two gates, are less stable than single-gate oscillators and are also prone to oscillation on the wrong mode if improperly designed. They have nevertheless been widely used, perhaps because of their (theoretically, at least) minimum number of external components. A dual-gate oscillator which has been found to work satisfactorily in some applications (see Figure 7-27) uses a series resonant crystal. It can be used up to about 20 MHz with TTL gates. In many applications this oscillator has been used without the series resonant circuit consisting of L_1 and C_2 between the two gates. The inductor L_1 is simply omitted and C_2 is replaced by a bypass capacitor. The series resonant circuit contributes nothing at

*See footnote p. 66.

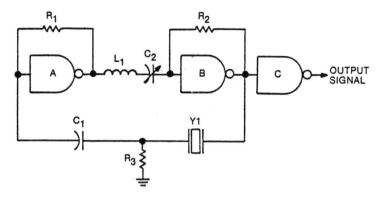

Figure 7-27. Basic dual-gate oscillator circuit.

the desired frequency of oscillation; rather, it provides protection against oscillation on the wrong mode. In some applications, particularly at low frequencies, R_3 and C_1 have been successfully omitted as well.

The two gates, U-A and U-B, are biased into the active region by resistors R_1 and R_2 to provide an amplifier with a gain A at a phase angle of approximately 360 degrees.

The feedback network is composed of the crystal, represented by an impedance $(R_e + jX_e)$, R_3, and C_1. At low frequencies the amplifier presents no phase shift, and C_1 is not required to correct for lagging phase shift in the gates. R_3 may be used to stabilize the input impedance of U-A and generally to present a lower resistance into which the crystal may work.

The oscillator may be analyzed by the use of equation (3-1) (Chapter 3) if the amplifier is represented by its Y-parameters and the feedback network by its Z-parameters. The Z-parameters may be written from inspection and are:

$$Z_{11} = (R_e + R_3) + jX_e \tag{7-72}$$

$$Z_{12} = Z_{21} = R_3 \tag{7-73}$$

$$Z_{22} = R_3 + jX_{C1} \tag{7-74}$$

where $X_{C1} = -1/\omega C_1$. \hfill (7-75)

Also defining $\Delta Z = Z_{11} Z_{22} - Z_{12} Z_{21}$, we have \hfill (7-76)

$$\Delta Z = (R_3 R_e - X_e X_{C1}) + j[R_3 X_e + (R_e + R_3) X_{C1}]. \tag{7-77}$$

The analysis may be considerably simplified by assuming that the input and output admittances of the amplifier are real and that the reverse transfer admittance is negligible. Then $y_{11} = g_{11}$, $y_{12} = 0$, $y_{21} = g_{21} + jb_{21}$, and $y_{22} = g_{22}$. Also defining $\Delta y = y_{11}y_{22} - y_{12}y_{21}$, we have $\Delta y = g_{11}g_{22}$.

The general equation for oscillation given in Chapter 3 requires that:

$$y_{21}Z_{21} + y_{11}Z_{22} + y_{22}Z_{11} + y_{12}Z_{12} + \Delta y \Delta Z + 1 = 0. \quad (7\text{-}78)$$

Substituting the values for this oscillator, we have

$$(g_{21} + jb_{21})R_3 + g_{11}(R_3 + jX_{C1}) + g_{22}[(R_e + R_3) + jX_e]$$

$$+ g_{11}g_{22}(R_3R_e - X_eX_{C1}) + jg_{11}g_{22}(R_3X_e + R_eX_{C1} + R_3X_{C1}) + 1 = 0.$$

$$(7\text{-}79)$$

The imaginary part of the equation is given by

$$X_e = \frac{-b_{21}R_3 - X_{C1}g_{11}(1 + g_{22}R_e + g_{22}R_3)}{g_{22}(1 + g_{11}R_3)}. \quad (7\text{-}80)$$

The value of X_{C1} required to operate the crystal at series resonance is found by setting $X_e = 0$ and solving for X_{C1}.

$$X_{C1} = \frac{-b_{21}R_3}{g_{11}[1 + g_{22}(R_e + R_3)]}. \quad (7\text{-}81)$$

In the low-frequency case, when $b_{21} = 0$, we see that $X_{C1} = 0$ and R_3 is unnecessary. If b_{21} is not zero, we see that since the product R_3b_{21} occurs in equation (7-80), smaller values of R_3 result in a more stable oscillator so long as the crystal drive level is not excessive.

At higher frequencies the required value of C_1 may be quite small, and this may lead to unstable operation because of other potential modes of operation. It may therefore be desirable to place a 20- to 30-pF capacitor directly in series with the crystal to bring it up to frequency rather than to make C_1 too small.

Referring now to equation (7-79), we see that the real terms result in the equation

$$\frac{g_{21}}{g_{22}} + \frac{g_{11}}{g_{22}} + \frac{R_e + R_3}{R_3} + g_{11}R_e - \frac{g_{11}X_eX_{C1}}{R_3} + \frac{1}{g_{22}R_3} = 0.$$

Now representing the open-loop voltage gain of the amplifier by $A = -g_{21}/g_{22}$ and assuming that C_1 is adjusted so that crystal reactance $X_e = 0$, we have

$$A \geqslant \left(\frac{R_e + R_3}{R_3}\right) + g_{11}\left(R_e + \frac{1}{g_{22}}\right) + \frac{1}{g_{22}R_3} \qquad (7\text{-}83)$$

for oscillation to start.

A few comments may be appropriate regarding the Y-parameters of the two gates in cascade. It is desirable simply to measure the admittance parameters of the amplifier. If, however, the Y-parameters of the individual gates are known, then it can be shown that the presence of a biasing (feedback) resistor modifies the Y-parameters according to the following relationships:

$$y_{11} = y_i + \frac{1}{R_1}, \qquad y_{12} = y_r - \frac{1}{R_1}$$

$$y_{21} = y_f - \frac{1}{R_1}, \qquad y_{22} = y_o + \frac{1}{R_1},$$

where y_i, y_r, y_f, and y_o are the Y-parameters of an individual gate. It can also be shown that placing two gates in cascade results in a final set of Y-parameters:

$$y_{11} = y_{11a} - \frac{y_{12a}y_{21a}}{y_{11b} + y_{22a}} \qquad (7\text{-}84)$$

$$y_{12} = -\frac{y_{12a}y_{12b}}{y_{11b} + y_{22a}} \qquad (7\text{-}85)$$

$$y_{21} = -\frac{y_{21a}y_{21b}}{y_{11a} + y_{22b}} \qquad (7\text{-}86)$$

$$y_{22} = y_{22b} - \frac{y_{12b}y_{21b}}{y_{11b} + y_{22a}}. \qquad (7\text{-}87)$$

As a practical matter, it may be satisfactory to experimentally optimize component values from typical gate oscillators in the same frequency range. Table 7-3 gives typical values for the components of gate oscillators at 7 MHz, 9 MHz, and 20 MHz.

As noted earlier, the frequency stability of the dual-gate oscillators

Table 7-3. Typical Values for Dual-Gate Oscillator.*
(See circuit diagram of Figure 7-27)

Frequency (MHz)	Gate type	R_1 (kΩ)	R_2 (kΩ)	R_3 (Ω)	C_1 (pF)	C_2 (pF)	L_1 (μH)
7	54LS04	1	3.9	none	none	1000	none
9	5404	0.680	0.680	100	470	20	15
20	54LS04	0.680	2.2	100	100	10	12

*See footnote p. 66.

is poor compared to the oscillators discussed previously. For the 20-MHz oscillator, typical gates pull the crystal about 5 ppm over a 70°C temperature range. This varies greatly from gate to gate and may be as high as 50 ppm. The frequency of the test oscillator also changed from 1 to 3 ppm for a 0.1-V supply voltage variation, which is about 2 orders of magnitude worse than for the single-gate circuit of Figure 7-26.

The 20-MHz crystal operates about 500 Hz below series resonance, which can be corrected by placing a 27-pF capacitor in series with it. The frequency can also be raised by making C_1 = 27 pF; however, with this value of C_1, the circuit is bordering on instability and must be carefully checked in the final mechanical configuration.

The voltage coefficient of the 7-MHz dual-gate circuit was found to be 4.6 ppm for a 0.1-V supply voltage change. The crystal operates about 2 kHz below series resonance.

It should be noted that the dual-gate oscillators, like the single-gate units, are prone to free-running oscillations, particularly if the crystal is not present; this must be considered in making the choice to use a gate oscillator.

Performance of the 9-MHz oscillator is much the same as that of the 20-MHz circuit. If a 54LS04 is used, the waveform is slightly improved by increasing R_2 to 2.2 kΩ.

7.7. INTEGRATED-CIRCUIT OSCILLATORS

A large number of integrated circuits (ICs) are available which can be used as crystal oscillators or which include a crystal oscillator. The information presented here regarding these circuits is related both to the design and the application of the devices, although it is somewhat slanted toward the application. Many existing ICs require only the attachment of an external crystal, while some require other components as well. The circuits at the time of this writing tend to fall into three categories. The first provides a single bipolar or field effect transistor to which the external crystal and feedback network can be attached. For this class of circuits the design equations developed for transistor oscillators earlier in the section are directly applicable, and the frequency stability is generally quite good.

A second class of circuits, often using MOS technology, provide a gate which can be used as a crystal oscillator. The design techniques developed in section 7.6 for gate oscillators are then directly applicable and the frequency stability is generally equivalent to that of oscillators using discrete gates of the same type.

The third class of circuits is designed with a multistage amplifier on the chip and the external crystal either closes the feedback path from the amplifier output to its input or it serves as a frequency-selective bypass at some point in the amplifier. Many of these circuits are used as clock drivers for microprocessors (or on microprocessors), as frequency synthesizers, modems, TV circuits, phase-locked loops, and the like. As might be expected, the frequency stability varies greatly with the design, and while some are good, others are very poor indeed.

Because of the large number of circuits being introduced and/or available, a detailed treatment of specific circuits is impractical. A number of general comments and principles apply, however, which are helpful. It should be obvious that in most cases the application of sound design principles will result in an oscillator of increased stability at essentially no difference in cost from a poorly designed circuit and may make the product more useful. For example, a microprocessor may use a crystal to stabilize the clock frequency on the chip. A frequency error of 1 percent may be almost inconsequential with respect to operation of the processor; however, if the oscillator is designed well, the inherent stability may be ±0.0025 percent or better. The clock can then be used as the reference oscillator to control the carrier of a radio transmitter, the time base for a digital clock, or some other function as required. In such equipments the applications engineer may wish to examine the stability of several circuits to find a suitable unit for his purpose.

Since a considerable variety of amplifier configurations is possible on a chip, no attempt is made here to analyze specific circuits. An analytic treatment is developed based on the terminal parameters of the circuits. Several general comments can also be made regarding the design of multistage integrated oscillators. Since an oscillator is sensitive to both the amplitude and phase of the amplifier, circuits with a considerable amount of phase shift will cause the crystal to operate well below (or in some cases above) series resonance. To operate on frequency the oscillator must be designed to require the same reactance for which the crystal was calibrated when manufactured (see Chapter 5). This can be determined by measuring the frequency of the crystal at series resonance (see Chapter 5), or the desired load point, and adjusting the oscillator components to obtain that frequency. If the phase shift is not too severe a series capacitor can often be used to raise the crystal frequency. Very small values of capacitance may indicate that the amplifier is unsuitable, and may also result in a tendency for the oscillator to free-run through the C_0 of the crystal (see Chapter 5) rather than at the piezoelectric resonance. If sufficient gain and phase shift are present, free running may take place through C_0 even though no external capacitor is used.

It should be noted that the crystal can oscillate on odd mechanical overtones as well as on the fundamental frequency. If the gain is higher at the overtone frequency than on the fundamental, and if no tuned circuit is used, oscillation on the overtone will result. Conversely, if operation on an overtone is desired, it will in general be necessary to provide a tuned circuit which limits the region of gain to the vicinity of the desired overtone.

If the oscillator being designed may be operated over a large range of frequencies, it is important to check it at all frequencies in the band to ensure, first, that oscillation will always occur, and secondly, that free-running oscillations will not occur. A 20-MHz crystal may have a resistance of 10–20 Ω, while a 100-kHz crystal may have a 100-kΩ series resistance. It does not follow that oscillation at 20 MHz guarantees oscillation with a 100 kHz crystal as well.

Extensional and flexural mode crystals in the low-frequency region may have active spurious responses near the desired response. Excess gain in the oscillator may in some cases result in oscillation on these spurious modes rather than on the desired frequency. When the circuit is turned on, oscillation builds up on all frequencies for which the phase requirement of 360 degrees occurs and the gain is greater than unity. The mode reaching the saturation amplitude first or having the most gain generally will survive and suppress the others (although it is possible to sustain multiple oscillations in some oscillators with high gain). The circuit should be carefully examined under as many conditions as possible to ensure that spurious oscillations will not occur. It is also good practice to check the frequency over the required temperature range as well as the frequency change due to supply voltage variation and load changes. The frequency drift over temperature caused by the IC can be determined approximately by connecting the crystal to the oscillator with the crystal external to the temperature chamber. A high-impedance balanced transmission line may be suitable for the connection. In some cases it may be necessary to use a small blower on the crystal to prevent temperature changes resulting from thermal conduction in the transmission line. A more accurate procedure is to measure the crystal temperature coefficient with a CI meter, Vector Voltmeter test set, or bridge, and subtract the crystal frequency drift from the total temperature coefficient. In a good oscillator the frequency change caused by the active circuit will be insignificant compared to that caused by the crystal.

While it is desirable in the design of integrated-circuit oscillators to use a set of analytic tools, the detailed equations for oscillation are generally too complex to be useful. Two approaches are presented here based on the terminal parameters of the integrated circuit. In those circuits where the crystal acts as a frequency selective bypass in the amplifier which is internally crosscoupled, it may be convenient to think of the circuit as a negative-resistance element in series with an inductance. An approximate equivalent circuit is shown in Figure 7-28. A series compensating capacitor C is shown in series with the crystal. For on-frequency operation with a series resonant crystal, C should be resonant with L_0 at the nominal frequency of the crystal. The resistance R_n is a negative value and must be larger in magnitude than the equivalent resistance of the crystal for oscillation to take place.

It is possible to determine the magnitude of R_n in several ways.

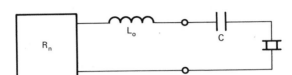

Figure 7-28. Negative-resistance oscillator with series compensating capacitor.

Perhaps the most obvious is to place a crystal between the appropriate terminals of the IC and add series resistance until oscillation will no longer occur. The magnitude of the negative resistance is then given by the sum of the crystal resistance and the additional series resistance. The magnitude of the oscillator inductance can be found by noting the difference between the frequency of oscillation and the series resonant frequency of the crystal (without C or the series resistance) and calculating*

$$L_0 = \frac{1}{(2\pi f_s)^2 \left(\dfrac{C_1}{2\Delta f/f} - C_0 \right)}. \tag{7-88}$$

It can also be found experimentally by selecting C to obtain the series resonant frequency of the crystal. Then

$$L_0 = \frac{1}{(2\pi f_s)^2 C}. \tag{7-89}$$

Since the equivalent inductance will in general vary as a function of frequency it should be computed near the nominal frequency of the crystal to be used.

It is desirable to minimize the equivalent inductance of an oscillator for several reasons. First, the equivalent inductance will change with temperature and supply voltage, causing the oscillator frequency to drift. Secondly, it may result in free-running oscillations through the C_0 of the crystal.

The equivalent inductance is a result of phase shift in the amplifier and can be minimized in the design by using as few stages as possible

*See Figure 5-1 for definition of terms.

and by increasing the bandwidth of the amplifier. The negative resistance will, of course, be a function of the gain of the amplifier and the impedance level where the crystal is placed.

In general the circuit may have a negative impedance characteristic such as that shown in Figure 7-29. The negative-resistance region is restricted to a portion of the voltage current characteristic. As oscillation builds up, the voltage swings beyond the negative-resistance region and the equivalent resistance becomes less negative. Finally at saturation (or equilibrium if AGC is used) the equivalent negative resistance equals the positive resistance of the crystal. It can be shown[13] that the crystal current builds up according to the equation:

$$i(t) = Ke^{-at} \sin(\omega t + \theta) \tag{7-90}$$

where

$a = (R_1 + R_n)/2L$, and

$\omega = \sqrt{(1/LC) - a^2}$.

Here L is the sum of the crystal inductance and L_0. C_0 was neglected in the analysis. So long as R_n is negative and larger than R_1, the amplitude continues to build up. Finally when $R_1 = -R_n$ an equilibrium condition is reached. Equation (7-90) also shows that the frequency of oscillation is lowest initially and increases slightly as the circuit stabilizes.

Test data on several ICs of the crosscoupled type shows a wide variation in equivalent inductance, from approximately 1–2 μH to greater than 250 μH over the frequency range from 1 to 20 MHz. Therefore, while some ICs operate with the crystal near series resonance, others operated as much as 1 percent low in frequency.

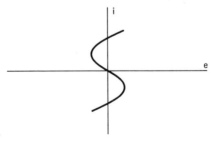

Figure 7-29. Negative-resistance characteristic of oscillator.

Some manufacturers provide excellent data sheets which not only characterize the equivalent circuit but also provide data on the environmental stability, and in some cases on the noise level as well. Others provide little more than a pin connection diagram showing where to connect the crystal. To achieve maximum versatility of the product, designers are encouraged to provide complete information on their oscillators. Conversely, if such information is not provided, the applications engineer should run complete tests on an IC prior to committing the circuit to production, particularly if a frequency stability approaching that of the crystal is required. Typical temperature coefficients for the better ICs of the cross-coupled type are 10–15 parts in $10^8/°C$ exclusive of the crystal temperature change.

As indicated earlier, a number of ICs use an amplifier on the chip and connect an external crystal between the output and input of the amplifier. While these circuits can be thought of as negative-resistance oscillators, it may be desirable to analyze them as amplifiers with a feedback network composed solely of the crystal. Such a circuit is shown in Figure 7-30. Here as before, the Y-parameters are used for the amplifier. From the definition of these parameters (see Appendix A) we may write the equations:

$$I = y_{11} V + y_{12} V' \tag{7-91}$$

$$-I = y_{21} V + y_{22} V'. \tag{7-92}$$

We also not that

$$V = V' - I(R_e + jX_e). \tag{7-93}$$

Solving the simultaneous equations for V gives:

$$V = \frac{\begin{vmatrix} 0 & -y_{12} & 1 \\ 0 & -y_{22} & -1 \\ 0 & -1 & R_e + jX_e \end{vmatrix}}{\begin{vmatrix} -y_{11} & -y_{12} & 1 \\ -y_{21} & -y_{22} & -1 \\ 1 & -1 & R_e + jX_e \end{vmatrix}}$$

Since the numerator is zero, V will be zero unless the denominator is

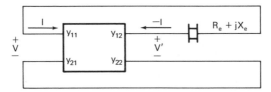

Figure 7-30. Integrated circuit with external crystal.

also zero. The conditions for oscillation may then be found by equating the determinant in the denominator to zero, which gives

$$\sum y + R_e \Delta y + jX_e \Delta y = 0, \qquad (7\text{-}94)$$

where

$$\sum y = y_{11} + y_{12} + y_{21} + y_{22}, \text{ and}$$

$$\Delta y = y_{11}y_{22} - y_{12}y_{22}$$

If there is no phase shift in the amplifier, the Y-parameters are real and from equation (7-94) we see that the imaginary part is satisfied by the condition $X_e = 0$, which is the desired operating point for a series resonant crystal.

Unfortunately, a multistage amplifier tends to have considerable phase shift, and the crystal normally operates below series resonance where X_e is negative. If the phase shift is not too severe, the frequency can be brought up to series resonance by placing a capacitor in series with it. (A note of caution here is that small capacitors, approaching the value of the holder capacitance C_0, may lead to free-running oscillations above the crystal frequency.)

An amplifier which has a considerable amount of phase shift at the operating frequency is undesirable from another consideration. The phase shift of any amplifier tends to change with temperature and voltage. An amplifier with a large phase shift will therefore also have a large phase change with temperature and supply voltage variation. The phase changes result in a change in frequency so that the crystal reactance makes up for the phase difference. The crystal reactance changes rapidly with frequency; however, variations of hundreds of parts per million can result particularly if the phase correction being made by the crystal is already approaching 90 degrees. It is apparent, therefore, that the fewer stages in the amplifier, provided the gain requirement can be met, the more stable the oscillator will be. Indeed the circuits using a single transistor with two external feedback capacitors in a π network with the crystal tend to make the best oscillators.

It is possible to plot equation (7-94) in two parts and in some cases obtain a better understanding of the crystal reactance required for oscillation. If a Smith chart is used the curve given by

$$Z = -\sum y/\Delta y$$

can first be plotted at the nominal crystal frequency as a function of input amplitude. A second curve representing $R_e + jX_e$ for the crystal as a function of frequency can then be added and the intersection represents the frequency and amplitude of oscillation.

The oscillator is not limited to operation in the resonant region of the crystal, and a search can also be made for spurious oscillations. This can be done by plotting Im Z as a function of frequency on a rectangular graph along with X_e, which will be a plot of C_0, the holder capacitance, except near piezoelectric resonances. Any intersection of the two reactance curves where Re Z is greater than R_e represents a potential frequency of oscillation and should be searched for during testing of the circuit.

A very serious problem may arise in using the Y-parameters here as equivalent admittances as a function of amplitude, particularly if the output stage of the amplifier performs the limiting function. Since the Y-parameters are measured with the input/output short-circuited, the real limiter will then not be measured. Under these conditions, the oscillator output voltage will normally be close to the peak-to-peak swing of the output amplifier.

Chapter 8 deals with tests which should be performed on any oscillator prior to committing it to production. Although some of the material presented here is duplicated in Chapter 8 it is felt that the designer or user of an integrated-circuit oscillator should be familiar with the tests recommended.

8
Preproduction Tests for Crystal Oscillators

Crystal oscillators are among the more critical electronic circuits and, as such, often experience difficulty in production. Certain tests can be run on the engineering models of an oscillator to assist in assuring that the circuit will not encounter trouble in production or, at least, that it will not require a major redesign.

Perhaps the most important test is to build three to five engineering models and make certain that none of them is marginal. Secondly, it is important that the oscillator circuit will have adequate reserve gain to function with maximum-resistance crystals. This test can be run conveniently by adding resistance in series with the crystal until the circuit will just oscillate. The total resistance (crystal plus the external resistor) must be higher than the maximum permissible resistance for the crystal type being used. In order to allow for variations in transistors and other components, it usually is desirable to have the circuit oscillate with two or three times the maximum crystal resistance.

A number of other factors should be considered in the evaluation of an oscillator. Several of these are obvious but are included here for completeness.

a. *Frequency stability.* The required frequency tolerance of the oscillator must be larger than the temperature variation of the crystal and oscillator plus the aging rate of the crystal. If no tuning adjustment is used, allowances must be made for unit-to-unit variation also.

b. *Output power.* The output power must be adequate even with a high-resistance crystal and a low-gain transistor.

c. *Crystal dissipation.* The crystal dissipation should be consistent with the discussion in section 5.4.

d. *Spurious oscillations.* The circuit should have a reasonable safety factor with respect to free-running and spurious oscillation. As a rule of thumb, the oscillator should be capable of being detuned to pull the crystal frequency at least ±10 ppm without spurious or free-running oscillations becoming evident. Switching the crystals and turning the oscillator off and on repeatedly in the detuned condition also may aid in discovering spurious oscillations. These procedures are inadequate, however. The only way to determine a spurious safety factor with any degree of certainty is with the aid of very low-spurious-ratio crystals for which spurious oscillation actually can be induced.

e. *Component tolerance.* The critical components of the circuit should be checked for permissible tolerance. This can be done conveniently by substituting components of the next size smaller and larger. In the case of gate oscillators, it is particularly important to obtain limit-active devices to check for spurious and free-running oscillations.

f. *Stray capacitance variation.* The circuit must be capable of functioning properly even if the stray capacitance or inductance configuration changes somewhat. This usually is unimportant for oscillators below 10 MHz. It can be checked by adding a 1- or 2-pF capacitor to ground at every critical point of the circuit.

g. *Crystal load capacitance.* The oscillator should be designed so that the crystal looks into the specified load capacitance (32 pF, 20 pF, or series resonance). This can be determined easily by having the frequency of a crystal measured at the desired load capacitance and then adjusting it to that frequency in the actual circuit.

h. *Supply voltage.* The oscillator should be checked to make certain that it will function properly with the minimum and maximum supply voltages.

i. *Temperature.* The oscillator should be checked over the entire temperature range with all conditions exactly the same as they will be in the final unit.

9
Other Topics

9.1. CRYSTAL SWITCHES

It is sometimes desirable to use a single oscillator with a number of crystals and a switching mechanism. This usually presents no problem at HF frequencies where the stray capacitance and inductance of the switch have negligible effects. At VHF frequencies, however, these strays can be extremely detrimental to the circuit performance.

There are two methods of lessening the problem and usually both are necessary in order to achieve satisfactory performance. First, an oscillator type which is reasonably tolerant of strays should be used. Second, every possible effort should be made to minimize the strays. The most effective means of accomplishing this is to build the crystal switch small. In the case of printed circuit switches, the use of small pads reduces capacitance. The switch collector rings often can be placed to minimize coupling to other parts. It may not be desirable to make the collector rings too narrow, however, as this adds to the stray inductance. It is very desirable to use a switch wafer which has a low dielectric constant. Teflon base materials are considerably superior to glass epoxy or phenolic boards in this respect. The bond strength is somewhat lower, however.

Depending on the oscillator type being used, stray capacitance in some locations may be more detrimental than in others. For example, in the impedance-inverting Pierce oscillator, stray capacitance from the base side of the crystal to ground is only of secondary importance, while capacitance across the crystal is very important. Lead inductance is not as critical for this oscillator as it is in the grounded-base oscillator. In general, the circuit should be examined to see which strays will cause the most harm; then they should be minimized, even at the expense of other strays.

A concept which is often forgotten in crystal switching is that unused crystals must be effectively removed from the circuit. If this is not accomplished, the unused crystals may absorb power from the oscillator should they have spurious responses near the frequency of the oscillating crystal. This may result in serious degradation of performance or even cause spurious frequencies in some cases. Perhaps the most effective method of eliminating coupling to the unused crystals is to short them to ground. If this cannot be done, then the stray capacitance to unused crystals must be kept very low.

Stray coupling is often quite severe if diode switching is used. The diodes themselves usually are quite good but switching schemes which save diodes usually are not good. Often, systems employ ingenious methods to switch a number of crystals with relatively few diodes. They usually have several sneak paths through the C_0's of the unused crystals which may cause serious trouble or even oscillation on the wrong crystal. It is, therefore, recommended that a thorough investigation of stray paths in a diode switch be made if fewer diodes than quartz crystals are being used. It is sometimes possible to tune out stray capacitance by the use of an inductor in parallel with it. This also may be done to resonate out sneak paths in diode switches.

Switching of crystals should be discouraged above 100 MHz since the oscillators themselves are very critical. A switch merely makes a bad situation even worse.

If crystal switching in the VHF range is necessary it may be desirable to examine the impedance-inverting Pierce oscillator, since it is less susceptible to stray inductance than the grounded-base oscillator.

The availability of frequency synthesizers on a single chip in certain frequency ranges (for example, the CB band) may well make the crystal switch obsolete within the near future, and the application of these devices should not be overlooked before making the decision to use a crystal bank with switches.

9.2. PULLABLE OSCILLATORS

It is sometimes necessary to vary the frequency of an oscillator by a small amount and yet require that the oscillator be quite stable. This can be done using a crystal oscillator provided the pullability requirements are not too severe. The frequency can usually be varied several hundred parts per million without much difficulty. If extreme measures are taken, the pullability may be as high as several thousand parts per million. Pullability and stability are opposing requirements, and even in a crystal oscillator the stability will suffer as the pullability is increased.

Perhaps the best way to pull a crystal oscillator is to put a voltage-variable capacitor in series with the crystal, an example of which is shown in Figure 9-1.

The pullability is determined primarily by two factors: the reactance–frequency slope of the crystal, and the reactance–voltage curve of the varactor. Since the frequency of oscillation will be near the crystal frequency, the other reactances in the oscillator circuit may, for practical purposes, be considered to be constant. From equation (5-3), which gives the antiresonant frequency of the crystal as $f_a = f_s[1 + (C_1/2C_0)]$, we see that the pullability of the crystal is larger if the C_0/C_1 ratio is small. A table of the C_0/C_1 ratio for some fundamental and overtone crystals appears in Table 9-1.

From this table, it is obvious that a fundamental crystal is most desirable. However, as the frequency is increased, the thickness of the crystal blank decreases and the unit becomes very fragile. The use of fundamental-mode AT-cut crystals is not recommended above 30 MHz.

The reactance–frequency curve of a quartz crystal is shown in Figure 5-3 and is fairly nonlinear because of the presence of the holder capacitance (see Figure 5-1). If an inductor is placed across the crystal to tune out the C_0, then the curve acquires another pole below series resonance and there is a relatively large region between

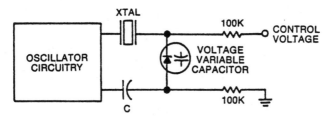

Figure 9-1. Typical pullable crystal oscillator: simplified diagram.

the two poles where the curve is linear. This slope is given approximately by:

$$\frac{dX}{df} = 4\pi L_1 \; \Omega/\text{Hz} \qquad (9\text{-}1)$$

where L_1 is in henrys.

Perhaps the best way to study the reactance–frequency curve for a crystal is to use the equivalent circuit of Figure 5-1, with an inductance in parallel with C_0, and to write a short computer program for the reactance. This was done for a 30-MHz crystal with the following parameters:

$R_1 = 20 \; \Omega$
$C_0 = 6 \; \text{pF}$
$C_1 = 0.03 \; \text{pF}$

The results are plotted in Figure 9-2 for the crystal alone and also for the crystal shunted by a 4.6-μH inductor. As can be seen, the curve is nearly a straight line. If a hyper-abrupt varactor with an exponent of unity is used in series with the crystal, a linear frequency–voltage curve results. Stray capacitance between the crystal and the varactor often reduces the pullability and linearity of a VCXO, and an advantage can be obtained by using two varactors in connection with series inductors. This arrangement, shown in Figure 9-3, allows

Table 9-1. Typical C_0/C_1 Ratios for Quartz Crystals.

Frequency (MHz)	Fundamental or overtone	C_0/C_1 ratio
0.2	fund.	400
2.0	fund.	270
6.9	fund.	230
8.8	fund.	220
12.5	fund.	200
31.0	third	2500
50.0	third	3000
60.0	third	3500
50.0	fifth	6200
60.0	fifth	6500

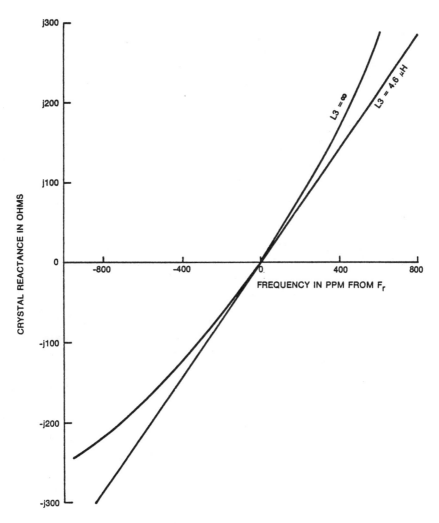

Figure 9-2. Reactance versus frequency for 30-MHz crystal.

the frequency to be pulled symmetrically above and below series resonance. A resistor is often necessary across the crystal to prevent free-running oscillations. Free-running oscillations are also minimized by using a grounded-base oscillator (see Chapter 7), since a tank circuit limits the range of oscillation to be near the crystal resonance.

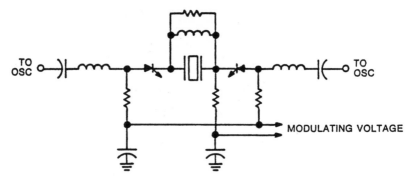

Figure 9-3. VCXO modulator schematic.

9.3. CRYSTAL OVENS

The design of crystal ovens is not considered in this book; however, it should be noted that when frequency stabilities beyond those obtainable by compensation are required, a crystal oven is often used. Crystal ovens are also used in some fixed-channel transmitters and receivers where replacing the crystal is required to change channels. A number of solid-state commercial crystal ovens are available for this purpose, and a stability in the order of 1 ppm can easily be obtained. The warmup time is usually not critical in these applications, and may be in the 5–15 min range. These ovens often accept plug-in HC-6/U or HC-35/U crystals and operate at 75°C or 85°C. If the warmup time can be tolerated, it may well be cost effective to use a solid-state oven designed to hold or clamp over the crystal.

In precision applications, both the crystal and the oscillator are usually packaged in the oven, and the oven is controlled by a thermistor sensor used in a bridge circuit with an operational amplifier. The amplifier drives a power transistor which controls the dc power to the oven heater.

A crystal oven of this type is capable of controlling the temperature of the crystal to within ±0.1°C over an ambient temperature range of −55°C to +75°C, and usually results in a temperature stability in the order of $\pm 1 \times 10^{-8}$. Various combinations of double ovens and hybrid arrangements using two heaters with a single control circuit are also available. The stability obtainable from a double oven is often in the $\pm 1 \times 10^{-10}$ region. Precision crystal ovens normally use foam insulation or a Dewar flask for insulation, while lower-precision units may use a dead air space. As discussed in section 4.1.1, crystal ovens have several major disadvantages which makes them unsuitable in some applications.

9.4. SQUEGGING, SQUELCHING, OR MOTORBOATING

Squegging is a term applied to relaxation-type oscillations which sometimes occur in addition to the desired mode of oscillation. In most instances the squegging rate is much less than that of the desired oscillation. Frequency differences of one or two orders of magnitude are not uncommon. If squegging is severe, it actually may start and stop the desired oscillation at its relatively slow repetition rate. In less severe cases, it merely modulates the desired signal. Squegging can be observed most easily using an oscilloscope with a sweep rate compatible with the relaxation oscillation. The squegging then will show up as a modulation envelope on the desired signal. It also may be observed on a spectrum analyzer or a receiver as sidebands on the oscillator output.

Squegging is generally the result of several conditions. One of the strongest influences on it is the ratio of capacitance from base to ground and emitter to ground. This results from the inability of emitter voltage to follow changes in the average base voltage when C_e is too large. When this occurs, the situation often can be corrected by decreasing the value of C_e, by increasing the base-to-ground capacitance or by moving the ground return of the feedback network to the emitter directly. Other conditions have an effect on squegging also, such as the collector load and the shape of the characteristic curves for the transistor. Some transistor types therefore have a greater tendency to squeg than others. As might be expected, the tendency of a transistor to squeg is dependent upon its operating point and, in some cases, changing the Q-point may eliminate the problem. Squegging may also occur in gate oscillators, and can usually be cured by changing the values of biasing components such as shunt resistors.

9.5. SPURIOUS OSCILLATIONS

Problems are encountered occasionally with spurious oscillations in crystal oscillators. In general, the problem results when the crystal has a low-resistance spurious mode and the oscillator becomes controlled by the spurious rather than by the main response. The problem can usually be eliminated by specifying a sufficiently high crystal spurious ratio (ratio of spurious resistance to main response resistance). From an economic standpoint, this may not always be the best solution, however, and it may be desirable to design the oscillator circuit so that it has minimum tendencies toward spurious oscillation. Several factors, including the choice of circuit type, have a considerable bearing on spurious operation and are summarized here.[13]

In general the antiresonant oscillator circuits (Pierce, Colpitts, and Clapp) are less likely to cause spurious oscillation. A crystal spurious ratio only slightly greater than unity usually will prevent spurious operation. The VHF series resonant oscillators are considerably more prone to spurious oscillation. A spurious-ratio specification in the range from 1.5 : 1 to 3 : 1 may well be required to prevent spurious oscillation.

Several "spurious-causing" phenomena have been isolated which, if present in a particular circuit, will enhance the ability of the circuit to oscillate on a spurious mode. These phenomena are as follows:

a. *Excessive tank circuit Q.* This makes the oscillator circuit frequency selective and, when the tank circuit is mistuned, it may discriminate against the main response and allow operation on the spurious response.

b. *Excessive loop gain.* The circuit should be designed to have the lowest possible loop gain commensurate with operation of high-resistance crystals under worst-case conditions.

c. *Circuit elements directly in series or in parallel with the crystal.* In circuits employing a C_0 compensation inductor, the value of the inductor should be as large as practical.

d. *Interaction with other crystals.* If a bank of crystals is to be used with a crystal switch, the unconnected crystals should be shorted out if possible.

e. *Switching to an active oscillator.* In general, the possibility of spurious oscillation is greater if a crystal is switched to an energized oscillator circuit than if the crystal is switched first and the circuit is then energized.

f. *Unequal initial excitation of main and spurious responses.* The presence of a parasitic or unintentional resonant circuit in the oscillator, which for a certain setting of the variable element is at the spurious frequency, may lead to spurious oscillation.

g. *Free-running oscillation.* Free-running oscillations or a circuit which can almost free-run may increase the possibility of spurious oscillation drastically if the free-running frequency is near the spurious-response frequency of the crystal.

In general it has been found, when analyzing spurious oscillations, that the mode which will survive is determined during the period prior to saturation when the oscillations are building up. All modes for which oscillation is possible begin to build up when the circuit is energized. The mode which builds up most rapidly causes the oscillator to saturate and the other modes to die out. To a good approximation, the oscillator may be considered linear before saturation and various modes can be analyzed independently (principle of superposition). This results in a considerable simplification if analytic treatments are to be considered in studying the behavior.

10

Temperature Compensation

The frequency stability of an AT-cut quartz crystal resonator as a function of temperature is determined primarily by the angle at which the resonator plate is cut from the mother quartz crystal. Curves showing this dependence are presented in Figure 5-6. These curves follow a cubic equation of the form

$$f = A_1(t - t_0) + A_2(t - t_0)^2 + A_3(t - t_0)^3, \qquad (10\text{-}1)$$

where f is the frequency difference between t_0, usually taken to be 20 or 25°C and the temperature t. (The values for the coefficients are given in section 10.4.) Improved frequency stability can be obtained by operating the units in a controlled-temperature environment such as a crystal oven; however, the disadvantages of such operation have become apparent in recent years, as discussed in sections 4.1.1 and 9.3. It is therefore desirable to develop other means for improving the frequency stability of quartz crystal resonators which are more nearly compatible with present-day requirements.

It has been known for many years that the resonant frequency of a crystal unit can be made to shift by placing a reactance in series with it. If this reactance is made to vary in such a manner that it counteracts the frequency shift of the resonator with temperature, a greatly improved temperature coefficient can be obtained. The advent of the varactor diode and the thermistor in the late 1950s first made this practical. A method of analog temperature compensation was developed in which a multiple thermistor–resistor network was used to generate the required voltage–temperature curve.[23]

At the time of this writing nearly all production temperature-compensated crystal oscillators (TCXOs) use this method, which we shall refer to as analog compensation. Generally, the frequency stability using this method can be made as good as 0.5 ppm from −55°C to +85°C in production by tailoring some elements in the thermistor network to the individual crystal being used. With great care, small numbers of units have been compensated to better than 0.1 ppm, but the procedure is rather tedious.

It is not surprising that, with the development of field programmable read-only memories (PROMs), digital compensation should be possible. Historically, because only small PROMs were available initially, the coarse compensation was done using analog networks, and the final corrections were made digitally. The general approach was to sense the temperature and use its value in digital form to address a memory. The contents of the particular memory location then contained the fine correction voltage required at that temperature. The digital correction voltage was converted to an analog signal and applied to a fine-compensation varactor in the oscillator. This technique is referred to as *hybrid analog–digital compensation* and makes frequency stabilities in the 0.1-ppm range practical in production. If a large memory is used so that the coarse analog compensation can be eliminated, we refer to the technique simply as *digital compensation*. Temperature compensation can also be accomplished by using microprocessing techniques in which the processor compensates its own clock oscillator or an external precision crystal oscillator.

These techniques are discussed in detail in the remainder of the chapter and, in some cases, experimental results are presented showing what can be achieved in practice.

10.1. ANALOG TEMPERATURE COMPENSATION*

As indicated earlier, most TCXOs in production at the time of this writing use analog techniques and, although many new designs will be digital, it is nevertheless of interest to discuss the technique. In general the procedure is to place a varactor in the oscillator circuit where it can pull the frequency at least as far as the crystal drifts in temperature. A voltage divider network composed of thermistors and resistors is then designed which will produce the required voltage–temperature function to compensate the oscillator.

---

*Several of the results presented in this section were developed under sponsorship of the US Army Electronics Command and are discussed in more detail in reference 1.

Normally, though not always, the varactor is placed in series with the crystal and has a capacitance–voltage function given by the equation

$$C = \frac{K}{(V + V_0)^n},$$

(10-2)

where K, V_0, and n are constants (actually these quantities are some-what temperature-dependent, but for a first approximation may be treated as constants). V_0 is the contact potential and is in the order of 0.75 V; n is primarily determined by the slope factor of the p–n junction and may be on the order of 0.3–2. For analog TCXOs, an n around 0.5 is often used and represents an abrupt p–n junction.

Solving equation (10-2) for V gives:

$$V = \left(\frac{K}{C}\right)^{1/n} - V_0.$$

(10-3)

The crystal load capacitance required to pull an amount $\Delta f/f_s$ in parts per million is given in equation (5-2) and may be arranged in the form

$$C_L = \frac{C_1}{2(\Delta f/f_s)} - C_0.$$

(10-4)

Substituting $\Delta f = f - f_s$ gives

$$C_L = \frac{C_1}{2\left(\dfrac{f - f_s}{f_s}\right)} - C_0.$$

(10-5)

Normally the crystal is cut slightly low in frequency so that when $C_L = 32$ pF the crystal is on frequency at 25°C. Thus we have $f = f_L$ at the load capacitance C_L. Now, defining Δf_L as $f - f_L$, the change in frequency from nominal, the load capacitance required by the crystal at Δf_L is given by

$$C_{x1} = \frac{C_1}{2\left[\dfrac{\Delta f_L}{f_s} + \dfrac{C_1}{2(C_0 + C_L)}\right]} - C_0.$$

(10-6)

The function C_{x1} can then be determined by substituting values of $\Delta f_L/f_s$ from the crystal curve such as shown in Figure 5-6. In practice only the limit values of C_{x1} are normally calculated at lowest and highest crystal frequencies. Then knowing the values of the oscillator capacitors, C_1 and C_2 (see Figure 10-11, for example), the varactor capacitance C can be calculated by recognizing that C_1, C_2, and the varactor are effectively in series. Thus

$$C = \frac{1}{\dfrac{1}{C_{x1}} - \dfrac{1}{C_1} - \dfrac{1}{C_2}}. \qquad (10\text{-}7)$$

Finally then, using equation (10-3), the required voltage extremes can be found.

A large variety of thermistor–resistor networks have been successfully used to generate the required voltage function for TCXOs. One such network which works well is shown in Figure 10-1. It can be shown that the transfer function for this network is given by:

$$\frac{V_o}{V_i} =$$

$$\frac{RT_3(R_1 + RT_1)(R_2 + RT_2)}{(R_1 + RT_1)(R_2 + RT_2)(R_3 + RT_3) + R_2 RT_2 [(R_1 + RT_1) + (R_3 + RT_3)]} \qquad (10\text{-}8)$$

$$RT_1(T) = RT_1(T_0)\,\exp\beta_1\left(\frac{1}{T} - \frac{1}{T_0}\right) \qquad (10\text{-}9)$$

$$RT_2(T) = RT_2(T_0)\,\exp\beta_2\left(\frac{1}{T} - \frac{1}{T_0}\right) \qquad (10\text{-}10)$$

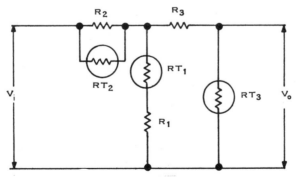

Figure 10-1. Three-stage thermistor network: schematic diagram.

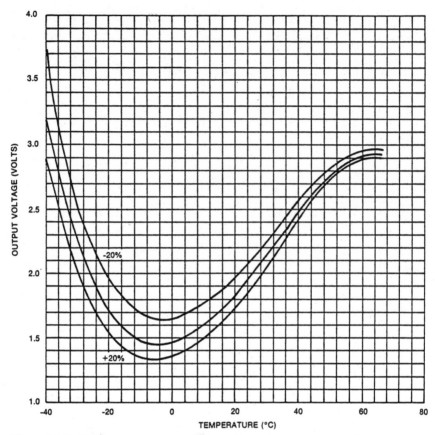

Figure 10-2. Voltage versus temperature for cold-temperature potentiometer R_2 (R_2 = 220 kΩ ± 20 percent).

$$RT_3(T) = RT_3(T_0) \exp \beta_3 \left(\frac{1}{T} - \frac{1}{T_0}\right). \qquad (10\text{-}11)^*$$

To assist the designer in manipulating the values of this network, a series of computer-generated plots is included showing how the various circuit values affect different portions of the temperature curve. These graphs are shown in Figures 10-2 through 10-10.

---

*Here β is referred to as the beta of the thermistor and is a measure of how fast the resistance decreases with increasing temperature. This temperature coefficient is determined by the composition of the thermistor during manufacture. T and T_0 are absolute temperatures in °K.

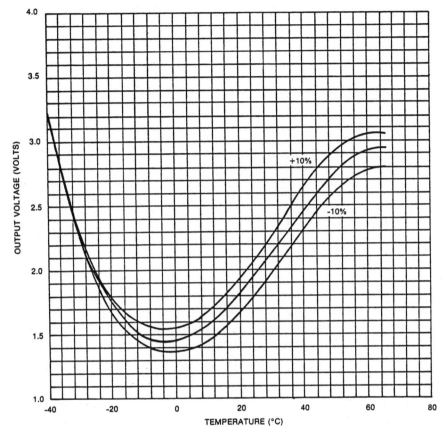

Figure 10-3. Voltage versus temperature for room-temperature potentiometer R_1 (R_1 = 22 kΩ ± 10 percent).

Because of the values chosen, the transfer function around room temperature is affected primarily by R_1 and RT_2, while the performance at cold temperatures is affected mostly by R_2 and RT_1. The transfer function at the high end of the temperature range is affected primarily by R_3 and RT_3.

A typical TCXO circuit diagram is given in Figure 10-11.

Because of the competitive nature of TCXO production, the actual procedures used by manufacturers to adjust the values of the thermistor network generally have not been available and the procedures in some cases involve as much art as science.

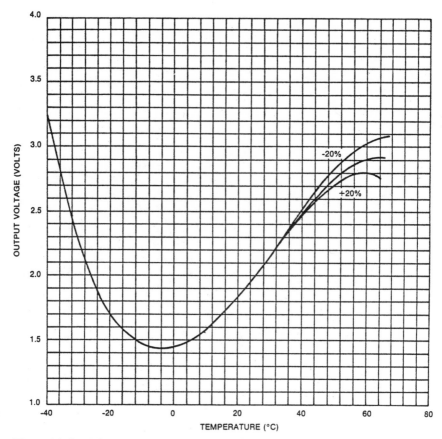

Figure 10-4. Voltage versus temperature for high-temperature potentiometer R_3 (R_3 = 33 kΩ ± 20 percent).

One approach which has been found to work well for the network of Figure 10-1 and an oscillator as shown in Figure 10-11 is described below.

The crystal is chosen to have an angle of cut so that the total frequency excursion between turning points is in the 35-ppm range. If an abrupt-junction varactor is used with an exponent of 0.5, its value is chosen to pull the crystal about 45 ppm from 1 V to two-thirds of the supply voltage. (A nominal value in the 20- to 50-pF range at 4 V dc should result.) A small selectable capacitor of perhaps 5 pF is

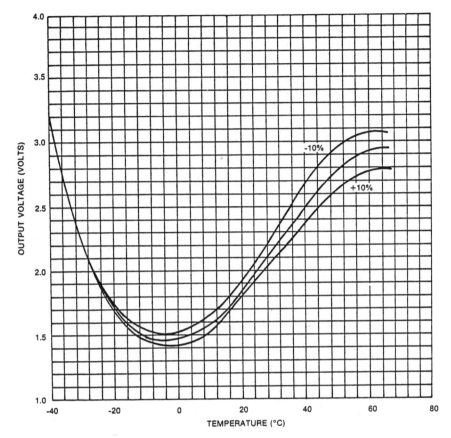

Figure 10.5. Voltage versus temperature for room-temperature thermistor RT_2 $[RT_2(T_0) = 100$ k$\Omega \pm 10$ percent, $\beta_2 = 3900]$.

retained across the varactor to allow final adjustment of the pullability. It has been found appropriate to use crystals which have a $\Delta f/f$ of approximately 160 ppm between series resonance and 32 pF.

For 12 V dc at the input of the thermistor network, R_1 (the room-temperature adjustment) is set for about 2.25 V initially. The oscillator is then placed in a temperature chamber and cooled to the lowest temperature, perhaps $-40°$C to $-55°$C, and R_2 (the cold adjustment) is set to put the oscillator back on frequency. The temperature chamber is then set to a temperature around the lower turning point of

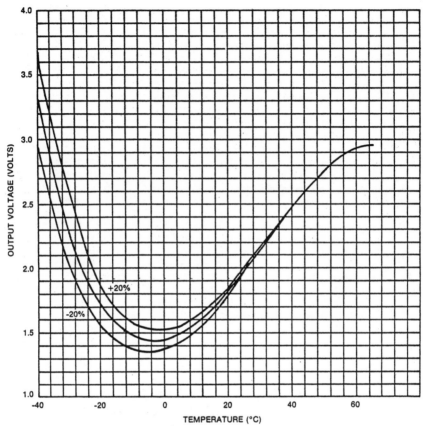

Figure 10.6. Voltage versus temperature for cold-temperature thermistor RT_1 [$RT_1(T_0)$ = 2.0 kΩ ± 20 percent, β_1 = 4410).

the crystal, usually about $-15°C$, and the frequency is checked. If the frequency is too high, the pullability is insufficient and the varactor shunt capacitor is reduced. If the frequency is too low, the shunt is increased. The procedure is then repeated, adjusting R_1 at $25°C$, and R_2 at the cold extreme until the region below room temperature is compensated. The temperature is then increased to the upper extreme, usually 75–$85°C$, and R_3 (the hot adjustment) is set to put the oscillator on frequency. A confirming temperature run is then made. A typical completed TCXO curve is shown in Figure 4-1.

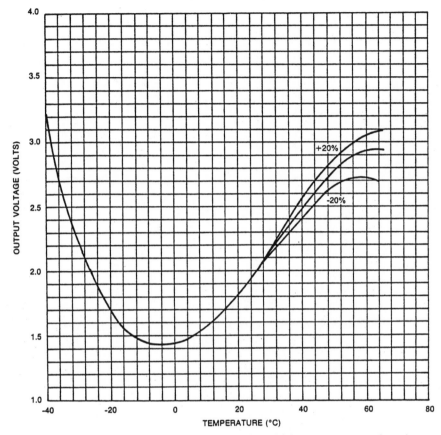

Figure 10-7. Voltage versus temperature for high-temperature thermistor RT_3 [$RT_3(T_0)$ = 1 MΩ ± 20 percent, β_3 = 5900).

For compensation to the 5- to 10-ppm range, a fixed compensation network with carefully specified parameters normally can be used, and the individual adjustments described above can be avoided. For tolerances in the 5- to 0.5-ppm range, individual adjustment is required. The compensation process can be reasonably automated for mass production.

Several other factors must be considered in the design of a TCXO, such as the voltage regulation at the input of the thermistor network and the load isolation at the buffer amplifier. Obviously, the voltage

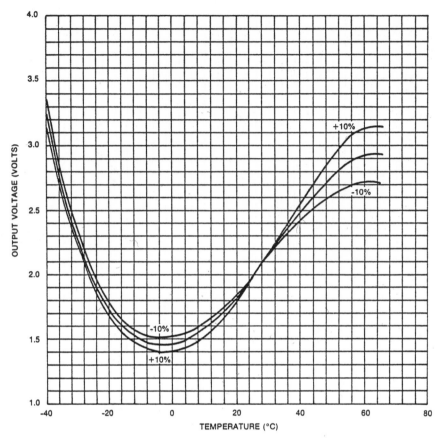

Figure 10-8. Voltage versus temperature for room-temperature thermistor RT_2 [$RT_2(T_0)$ = 100 kΩ, β_2 = 3900 ±10 percent).

regulation must be sufficiently good so that changes in the supply voltage will not cause the varactor to pull the frequency by a significant amount compared to the frequency stability specification. In general, either a Zener diode regulator or a packaged integrated circuit regulator is used to supply current to both the oscillator and the thermistor network. It is normally of considerable importance to minimize the power dissipated in a TCXO because of self-heating and temperature gradients which may cause a frequency drift at turn-on. Therefore, a low-power voltage regulator is recommended to keep the input power below the 100-mW range.

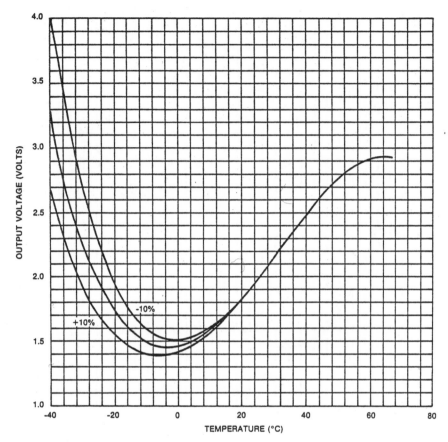

Figure 10-9. Voltage versus temperature for cold-temperature thermistor RT_1 [$RT_1(T_0)$ = 2 kΩ, β_1 = 4410 ± 10 percent).

As with changes in supply voltage, changes in the oscillator load can also perturb the frequency.

To analyze the effect, consider the phasor diagram of Figure 10-12, where E_s represents the signal voltage at some point in the oscillator loop and E_n represents an induced voltage from the output stage at the same point in the oscillator loop. Here the magnitude of E_n is exaggerated for purposes of illustration. The resultant voltage is shown as E_r. The phase shift in oscillator voltage caused by the presence of E_n is given by angle β. Angle α is the phase difference between E_s and E_n. By elementary geometry, we have $\overline{AB}$ = E_n and $\angle OAB$ = 180 - α.

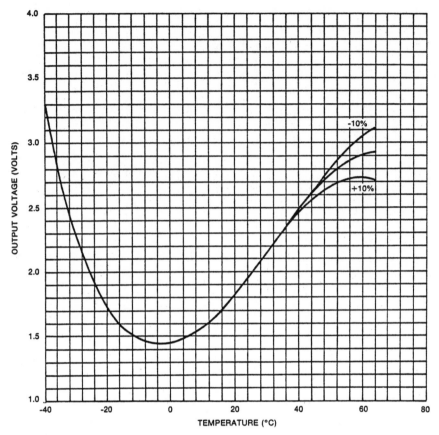

Figure 10-10. Voltage versus temperature for high-temperature thermistor RT_3 [$RT_3(T_0) = 1$ MΩ, $\beta_3 = 5900 \pm 10$ percent).

Then, by the law of sines, we have

$$\frac{\sin \beta}{\overline{AB}} = \frac{\sin (180 - \alpha)}{E_r}. \tag{10-12}$$

Rewriting and substituting E_n for $\overline{AB}$ gives

$$\sin \beta = \frac{E_n}{E_r} \sin (180 - \alpha) = \frac{E_n}{E_r} \sin \alpha. \tag{10-13}$$

If $E_n \ll E_s$ (as would be the case in an oscillator of practical in-

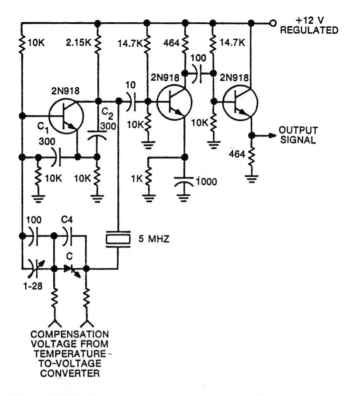

Figure 10-11. Temperature-compensated crystal oscillator and isolation amplifier.*

terest) we have $E_r \doteq E_s$ and $\sin \beta \doteq \beta$. Then

$$\beta \doteq \frac{E_n}{E_s} \sin \alpha. \tag{10-14}$$

It is now interesting to calculate the resulting frequency shift due to the presence of the induced voltage, E_n. To accomplish this, consider the circuit diagram of Figure 10-13. This is the circuit diagram of a crystal near series resonance. The current is given by

$$I = \frac{E}{R + j[\omega L - (1/\omega C)]}. \tag{10-15}$$

*See footnote p. 66.

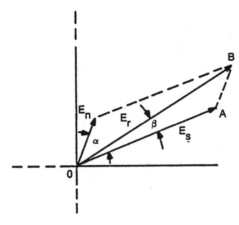

Figure 10.12. Oscillator voltage phasor diagram.

The phase angle of the current is given by

$$\theta = -\tan^{-1} \frac{\omega L - (1/\omega C)}{R}. \tag{10-16}$$

Differentiating with respect to ω, we have

$$d\theta = -\frac{1}{1 + \left[\dfrac{\omega L - (1/\omega C)}{R}\right]^2} \left\{\left[\frac{L}{R} + \frac{1}{\omega^2 RC}\right] d\omega.\right\}$$

Multiplying numerator and denominator in the second term on the

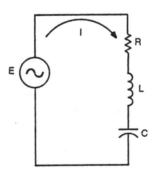

Figure 10.13. Series resonator circuit diagram.

right-hand side by $1/\omega$ gives

$$d\theta = -\frac{1}{1 + \left[\dfrac{\omega L - (1/\omega C)}{R}\right]^2} \left[\frac{\omega L}{R} + \frac{1}{\omega R C}\right] \frac{d\omega}{\omega}. \qquad (10\text{-}17)$$

If the circuit is near series resonance, then

$$1 >> \left[\frac{\omega L - (1/\omega C)}{R}\right] \quad \text{and} \quad \frac{\omega L}{R} \doteq \frac{1}{\omega R C} = Q.$$

Then

$$d\theta = -2Q\frac{d\omega}{\omega} = -2Q\frac{df}{f}. \qquad (10\text{-}18)$$

For small changes, $\Delta\theta = d\theta$ and we may write

$$\Delta\theta = -2Q\frac{\Delta f}{f}. \qquad (10\text{-}19)$$

We now consider an example to obtain a feel for the required magnitude of isolation. Suppose we desire $\Delta f/f$ to be less than 5 parts in 10^8 for a particular oscillator. Assuming a loaded Q of 25,000, we have

$$\Delta\theta = -2 \times 25{,}000 \times 5 \times 10^{-8} = -2.5 \times 10^{-3} \text{ rad.}$$

Substituting in equation (10-14), we find

$$\frac{E_n}{E_s} \sin \alpha = -2.5 \times 10^{-3}.$$

If α, the phase angle between E_s and E_n, is on the order of 90 or 270 degrees, the ratio E_n/E_s must be approximately 2.5×10^{-3} or, in decibels, E_n must be 52 dB below the oscillator voltage. If the voltage gain of the amplifiers is on the order of 20 dB, the required isolation is 72 dB. This assumes the worst case where E_n and E_s are at 90 degrees. If the phase shift of the amplifier chain is designed so that it is near zero or 180 degrees, much larger values of E_n can be tolerated. If, however, the load is allowed to vary from pure inductance to pure capacitance, α will vary approximately from -45 to $+45$ degrees, and the improvement would be only a factor of 0.707.

Still another consideration in the design of a successful TCXO is the elimination of hysteresis. If the oscillator frequency does not repeat exactly from temperature run to temperature run, it is nearly impossible to achieve successful compensation. Hysteresis is generally caused by a component such as a capacitor or resistor which does not repeat with temperature or which jumps slightly in value. For this reason, film resistors and monolythic ceramic capacitors are often used in TCXOs. If it is noted that the frequency of a TCXO does not repeat, it is usually wise to search for the faulty component prior to any further attempts at compensation.

Apparent hysteresis can also result from insufficient stabilization time at each temperature. The wise engineer will make sure that the frequency has truly stabilized prior to moving on to the next temperature. In many TCXOs, stabilization times in the 15- to 30-minute range are not uncommon.

In the discussion so far we have assumed that a varactor is used to pull the crystal frequency. It is also possible to achieve compensation in the 5- to 10-ppm stability range by placing thermistors directly in parallel and in series with fixed capacitors in the oscillator circuit. Although this approach is perhaps somewhat less elegant than the varactor approach, it is nevertheless cost effective in some applications. A particular advantage is the fact that a voltage regulator and varactor are not required.

10.2. HYBRID ANALOG-DIGITAL COMPENSATION[7, 57]

Analog temperature-compensated crystal oscillators, as described in section 10.1, with stabilities of ±5 parts in 10^7 from $-40°C$ to $+70°C$, have been a commercial reality for about a decade. Such units have been produced by the thousands at reasonable cost. Stabilities of 1–2 parts in 10^7 have been achieved in small quantities over the $-40°C$ to $+70-80°C$ temperature range, and stabilities of 5 parts in 10^8 have also been achieved over narrower ranges in quite small quantities. In general, compensation becomes increasingly difficult beyond ±5 parts in 10^7 because of the very small component tolerances involved, of the interaction of network adjustments, and of an undefined degree of electrical hysteresis in crystals due to thermal cycling. Partial solutions to these limitations, although not entirely desirable from a production standpoint, have been the use of digital computers to solve network calculations and the use of analog segmented networks to provide greater independence of adjustments. Often a large number of temperature runs have been required to "massage" units into the 1-2 parts in 10^7 stability region. For the computerized approach, this is due to a lack of accurate component data and the inability to install the exact component values calculated. For the analog segmented approaches, the major difficulties remain the lack of true independence between segments and the accuracy with which components must be selected.

The temperature-compensation method described in this section effectively eliminates the component accuracy problem and the interaction of segments while allowing better visibility to evaluate and minimize electrical hysteresis. Using this approach, it has been possible to achieve temperature compensation to ±5 parts in 10^8 over the $-40°C$ to $+80°C$ temperature range under controlled test conditions.

A major portion of the effort to develop this approach was carried out by G. Buroker under the sponsorship of the Solid-State and Frequency Control Division of the United States Army Electronics Command.

A block diagram of the TCXO is shown in Figure 10-14. Compensation is achieved with both analog and digital techniques. The analog, or coarse, circuit is used in a conventional manner to reduce the oscillator temperature coefficient (TC) to ±4 parts in 10^7 or less; then the digital network adds fine corrections to reduce the overall TC to less than ±5 parts in 10^8. The TCXOs with analog compensation need only minor design refinements to be used with the hybrid approach. Primarily, these are improved voltage and load coefficients and a reduction in power dissipation.

For a typical unit, the RF circuit consists of an oscillator followed by an isolation amplifier. Stages may be stacked in pairs to save power. Power consumption by the oscillator stage and the first buffer may be less than 5 mW from the regulated source. The second buffer may require two or three times more power to meet the output requirement; however, in any event, the power consumption which results in internal heating should be minimized.

The coarse compensation network consists of the three thermistors and their associated resistors, as described in section 10.1.

Separate varactors are recommended for coarse and fine compen-

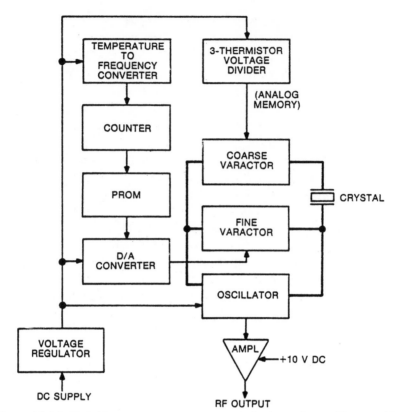

Figure 10.14. Hybrid   temperature-compensated   crystal   oscillator: block diagram.

sation, so that the size of fine correction steps can be held nearly the same at all temperatures.

A block diagram of the fine compensation circuitry is shown in Figure 10-15. The object of this circuitry is to correct errors greater than the stability specification that remain after coarse compensation. This is accomplished with the programmable memory that remembers the proper, independent correction voltages to apply at regular intervals in the temperature range.

For purposes of this discussion, a ±5 parts in 10^8 frequency stability specification over the -40°C to +80°C temperature range is assumed. This is about the limit of what can be achieved due to hysteresis effects in the crystal and other oscillator components. The hybrid

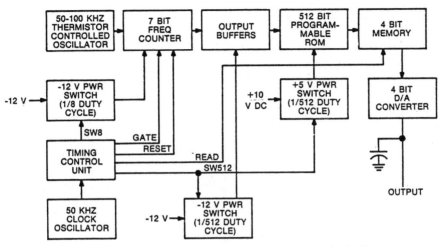

Figure 10.15. Digital fine-compensation network: block diagram.

approach can, of course, be used effectively in the 1–2 parts in 10^7 range but, as a practical matter, either straight analog or straight digital compensation (discussed in section 10.3) can be more economically employed for lesser stabilities.

Most of the development on the hybrid approach was accomplished during the early 1970s when the availability of PROMs and A/D converters in small sizes was relatively limited. Therefore, instead of using a temperature sensor followed by an A/D converter as described in section 10.3, a temperature-sensitive RC oscillator was used as the sensor, and the frequency counter as a means of converting the temperature to digital form.

The PROM was then addressed by the states of the counter. The D/A circuit then converts the output words of the programmed PROM to an analog voltage for TCXO correction. A clock and associated timing logic provide periodic updates and in general regulate the sequential operation of the counter.

Because of the relatively high power consumption of the PROM and the counters, switches are provided to sequence them on only when they are actually used; the control signals for these switches are generated by the timing control unit.

The required programming for the PROM is determined by stabilizing the TCXO at fixed temperatures, recording the states of the temperature registers, and simulating the correction output word with a manual switch. Required programming at intermediate temperatures is interpolated, and the entire PROM is then programmed and installed in the TCXO.

The timing signals for the digital circuitry are shown in Figure 10-16. These signals were developed in the custom MOS chip called DIGITCXO.* The first event, at the beginning of a cycle, is to energize the frequency counter flip-flops. This is accomplished by SW8 and the – 12-V switch shown in Figure 10-15. Once this is accomplished, the counter flip-flops (on the chip) are reset by a 20-μs pulse. The counter gate is then opened for approximately 2 ms. Upon completion of the count, the PROM is energized by SW512, and the latch is pulsed to store the outputs of the ROM. All circuits, except the timing control, are then deenergized for 18 ms. The DIGITCXO chip contains the 7-bit frequency counter and timing control unit. The finished chip measures 0.125×0.145 inch and was fabricated and packaged in a 22-pin, 0.5-inch round ceramic package by the Collins MOS facility at Newport Beach, California.

Compensation of the TCXOs is accomplished with the aid of an interface adapter which replaces the ROM during compensation. A 16-position rotary switch with binary format substitutes for the PROM output lines. The adapter allows the operator to stop, hold, and read an address and update by depressing a switch. The memory location being addressed is displayed in a decimal format using a 3-digit display.

Coarse compensation is accomplished with the ROM simulator switch set at midrange and is carried out using conventional techniques. The coarse network compensates the TCXO to ±4 parts in 10^7 from – 40°C to +80°C.

After the coarse compensation has been completed, the following steps are used for fine compensation:

    a. Seal the cover on the coarse portion of the TCXO in preparation for fine compensation.
    b. Stabilize the unit at room temperature for a minimum of

*Developed under sponsorship of the Solid-State and Frequency Control Division of the Electronics Components Laboratory, United States Army Electronics Command, Contract No. DAAB07-71-C-0136.

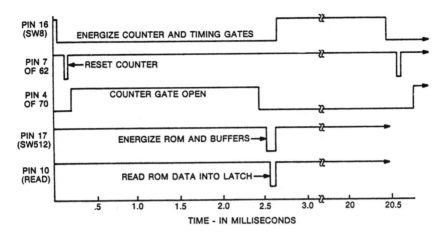

Figure 10.16. Timing diagram for MOS chip DIGITCXO.

10 hours. Beginning at -40°C, stabilize the sealed, coarsely compensated TCXO in 4-5°C intervals up to +80°C. At each temperature, record the PROM address and the decimal number of the (simulated) ROM output word that produces the smallest frequency error.

c. Tabulate the recorded PROM addresses and the desired corresponding output words. Estimate the required output words for intermediate PROM addresses by linear interpolation. (That is, if output words 8 and 3 were found to be required at addresses 107 and 112, respective intermediate interpolations are: 7 at 108, 6 at 109, 5 at 110, and 4 at 111.)

d. Program a PROM with the desired information using a PROM programmer.

e. Remove the cabling harness from the TCXO and install the programmed PROM. Clean the circuit boards and postcoat. Attach cover over digital compensation boards. (Alternatively, a confirming temperature run may be made before postcoating.)

f. Stabilize unit at room temperature for at least 10 hours. Repeat the preceding -40°C to +80°C temperature run at the same 4- or 5-degree increments to verify satisfactory performance.

The final frequency-temperature characteristic of a typical complete TCXO is graphed in Figure 10-17, along with the frequency-

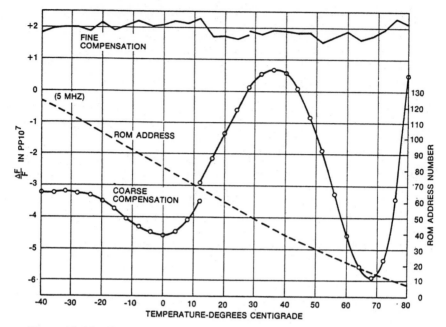

Figure 10.17. Characteristics of SN 8 after coarse and fine compensation.

temperature characteristic of the unit after coarse compensation. For this oscillator, the total frequency deviation was 70 parts in 10^8 before fine compensation and 7.5 parts in 10^8 after. Note that the curve has a discontinuity in midrange, caused by stopping the temperature run overnight. Stabilizing every 4°C from -40°C to +30°C, a complete temperature run requires two days. Also in the same figure is a plot of the PROM address versus temperature, indicating that a reasonably linear relationship was obtained.

10.3. DIGITAL TEMPERATURE COMPENSATION

The advent of larger PROMs and integrated A/D converters has simplified the compensation process so that TCXOs using entirely digital compensation are practical. The block diagram of such a unit is shown in Figure 10-18.

The crystal oscillator contains a single varactor, as in the case of the analog-compensated oscillators and, by application of the proper

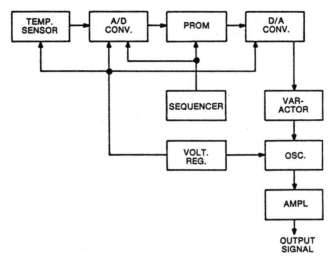

Figure 10-18. Digitally temperature compensated oscillator.

voltage to this varactor, the frequency is pulled by exactly the amount required to compensate for the temperature drift. In digital TCXOs it is generally desirable to use a hyper-abrupt varactor, with an exponent near unity, which gives a nearly linear frequency–voltage curve. The voltage range required can be found by the method outlined in section 10.1 with the aid of equations (10-3), (10-6), and (10-7). It should be noted that a voltage of less than about 1 V should be avoided because the RF voltage across the varactor may cause rectification and override the correction voltage.

The actual correction voltage is obtained in the following manner. First the voltage from the temperature sensor, such as a thermistor or diode, is digitized using the A/D converter. The digitized temperature word is then used as an address for the memory which contains the correction voltage required for the oscillator at that particular temperature. The contents of the memory location are latched into a digital-to-analog converter and the analog voltage is applied to the varactor. Since temperature changes relatively slowly, continuous corrections are not required. The temperature of an oscillator rarely changes more than 10 degrees per minute, and since the maximum rate of frequency change is less than 1 ppm/°C, a few corrections per second are sufficient. Consequently the A/D converter and the PROM can be turned off most of the time to save power and may be pulsed on only momentarily when a new correction is being made.

In most cases, it is desirable to minimize the memory required by the TCXO; thus it is important to choose the optimum word size for a given stability. It is most convenient to use uniformly spaced temperature intervals to address the memory. The slope of the frequency–temperature curve of a crystal, of course, varies greatly over the temperature range, as can be seen from Figure 5-6. The memory word size and capacity must be adequate to accommodate the worst-case slope. For a typical TCXO crystal with a frequency difference between the upper and lower turning points of about 32 ppm, we find slopes on the order of 1.4 ppm/°C at $-55°C$, -0.5 ppm/°C at 25°C, and 0.7 ppm/°C at 85°C. Let the worst-case slope be represented by S. Figure 10-19 shows how the compensation varies over a temperature interval from t_1 to t_2 if the exact compensation values are used at t_1 and t_2. The worst-case error occurs just before t_2 is reached when the frequency correction contained in the t_1 address is still being used but the crystal requires the value near t_2. This error can be cut essentially in half by overcompensating at t_1 so that the compensation is correct midway between t_1 and t_2. This is shown graphically between t_k and t_{k+1} in Figure 10-19. The frequency error due to the slope is then given by:

$$\Delta f_1 = \frac{(t_{k+1} - t_k)\,S}{2}. \tag{10-20}$$

Since a finite memory word size is used, an additional error occurs due to the fact that the exact desired value cannot always be obtained with a finite word size. In the worst case, the frequency can be set only to within one-half the frequency change represented by the least significant bit of the memory. Let this error be Δf_2. Then

$$\Delta f_2 = \frac{\text{total frequency correction range}}{2\,(\text{no. of correction levels})} \tag{10-21}$$

and the total worst-case frequency error is given by

$$\Delta f = \Delta f_1 + \Delta f_2. \tag{10-22}$$

We may then write:

$$\Delta f = \frac{S}{2}\left(\frac{T}{n}\right) + \frac{F}{2}\left(\frac{1}{2^b}\right),$$ (10-23)

where

S = maximum frequency–temperature slope of the crystal, in parts per million per Celsius degree.

T = the total temperature range over which the oscillator must operate, in Celsius degrees.

n = the number of words in the memory. Then each temperature interval is given by $t_{k+1} - t_k = T/n$.

F = the maximum peak-to-peak frequency excursion of the crystal, in parts per million.

b = the number of bits in each correction word.

Δf = the maximum frequency error to be allowed; in parts per million.

We wish to minimize the total number of memory bits required given by

$$M = nb.$$ (10-24)

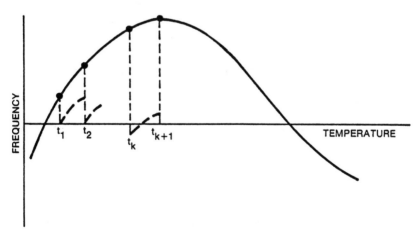

Figure 10-19. Frequency–temperature curve with exact compensation at t_1 and t_2.

Solving equation (10-23) for n gives

$$n = \frac{ST}{2\Delta f - (F/2^b)}. \tag{10-25}$$

Substituting equation (10-25) into equation (10-24) gives

$$M = \frac{bST}{2\Delta f - (F/2^b)}. \tag{10-26}$$

We may determine the value of b to minimize M by differentiating and setting $dM/db = 0$. We find

$$\frac{dM}{db} = \frac{\left(2\Delta f - \dfrac{F}{2^b}\right) ST - 2^{-b} bSTF \ln 2}{\left(2\Delta f - \dfrac{F}{2^b}\right)^2}. \tag{10-27}$$

Equation (10-27) is zero if the numerator is zero. Setting the numerator to zero leads to the expression

$$2^b = \frac{F(1 + b \ln 2)}{2\Delta f}. \tag{10-28}$$

Equation (10-28) cannot be solved in closed form; it must be evaluated by trial and error. Once b, the word size, has been determined, the number of words required can then be found from equation (10-25).

As an example, suppose that the following values are assumed:

$\Delta f = 0.1$ ppm,
$F = 32$ ppm,
$T = 85°C - (-55°C) = 140°C$, and
$S = 1.4$ ppm/°C.

Then from equation (10-28) we find

$$b \doteq 10.35 \quad \text{for} \quad \Delta f = 0.1 \text{ ppm}.$$

For $\Delta f = 0.5$ and 3 ppm, the values of b are 7.65 and 4.45, respectively. The values of M from equation (10-26) are given in Table 10-1 for several values of b and frequency stabilities of 0.1, 0.5, and 3 ppm for the crystal and temperature range above. Thus we see that it is possible to build a 0.5-ppm TCXO with an 8 × 256 PROM. In

Table 10-1. Memory Size for F = 32 ppm,
S = 1.4 ppm, and T = 140°C.

Word size (bits)	Memory size (bits)		
	0.1 ppm	0.5 ppm	3 ppm
4	–	–	196
5	–	–	196
6	–	–	214
7	–	1,829	238
8	20,906	1,792	267
10	11,615	2,023	–
11	11,694	–	–
12	12,238	–	–

many cases it may be convenient not to offset the corrections as was done at t_k and t_{k+1} in Figure 10-19 to achieve optimum performance. If compensation is accomplished simply by using the nearest available value at the temperature breakpoints, then twice the number of words are required.*

10.4. TEMPERATURE COMPENSATION WITH MICROPROCESSORS

From section 10.3 it can be seen that the memory requirements for digital compensation beyond 0.5 ppm are substantial and that it is desirable to operate on the stored data in some manner to reduce the number of correction values required. Perhaps the simplest algorithm which can be used is to interpolate between stored data points. This can be accomplished in several ways using digital logic or digital–analog combinations. Perhaps the most attractive means, however, is by the use of a microprocessor. Because of the availability of low-cost microprocessors, many items of communications equipment are being designed with a processor. In some cases, the microprocessor can be used to generate frequency corrections during idle time. A one-shot multivibrator can be used to request an interrupt every few seconds, or the processor may be programmed to service the TCXO

---

*Digital compensation of crystal oscillators is discussed in reference 47, which also contains several interesting variations of the basic approach discussed here.

at regular intervals. In other applications, as minimum microprocessor systems become available with self-contained I/O, PROM, and RAM on the chip, it is desirable to include a dedicated microprocessor in a semiprecision or precision frequency standard. The processor can be pulsed on momentarily to generate a correction and then turned off to save power and reduce self-heating.

An experiment was conducted to demonstrate the feasibility of temperature-compensating a crystal oscillator using the INTEL-8080 processor. A linear interpolation program was used to generate correction voltages from the following equation:

$$V = V_n + \frac{(V_{n+1} - V_n)(t - t_n)}{(t_{n+1} - t_n)}. \tag{10-29}$$

Here the temperature t is assumed to be between the stored values t_n and t_{n+1}, which correspond to compensating voltages of V_n and V_{n+1}, respectively.

A linear temperature sensor was used in the crystal oscillator and the output of the sensor was converted to an 8-bit digital number using a single-chip A/D converter. After the correction voltage was calculated, the output was converted to an analog signal and applied to the varactor. The program was written to allow 16-bit temperature data, although only 8 bits were used for the test oscillator.* It should be noted that although 16-bit temperature words can be accepted, the difference between any two adjacent temperatures may not exceed 7 bits.

A flow chart of the temperature-compensation program is given in Figure 10-20. Once the temperature is determined, a search is made for a correction value. If the exact value is found, no calculation is required and a test is made to determine if the correction voltage is the same as that determined on the previous pass. If so, the output value is left unchanged and a branch is made to repeat the program. If the correction value is different, the new value is read into the output latch and a branch is executed to the beginning of the pro-

---

*It is also possible to determine the temperature by using the processor to count the frequency of a thermistor-controlled *RC* oscillator. Approximately 33 machine level instructions are required to determine temperature in this way. The method was not used in this experiment.

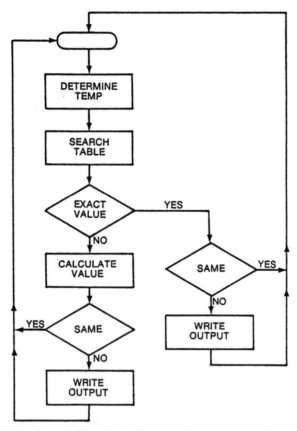

Figure 10-20. Overall flow chart of temperature compensation program.

gram. In the more likely event that the exact correction value is not stored for the particular ambient temperature under consideration, the search routine finds the nearest temperature–voltage pair above and below the actual temperature. The interpolation program then calculates a correction voltage based on equation (10-29), and a test is made to determine if the correction is different from the value found in the previous pass. If so, the output latch is updated and control is passed to the beginning of the program. If the output is the same, the latch is left unchanged.

A graph of the curve for the uncompensated crystal oscillator is shown in Figure 10-21 along with the final compensated curve. The

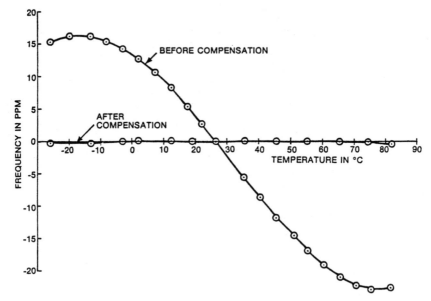

Figure 10-21. Frequency–Temperature for microprocessor compensated crystal oscillator.

compensated curve is also shown in Figure 10-22 with the ordinate expanded by a factor of 25. As can be seen, the frequency of this 5-MHz crystal oscillator is within $+2 \times 10^{-7}$ to -4×10^{-7} of the normal frequency over the entire temperature range from $-26°C$ to $+82°C$.* A schematic diagram of the oscillator is shown in Figure 10-23. The A/D converter, not shown, was an MM4357. The D/A converter, shown in Figure 10-24, consists of 8 CMOS buffers followed by a ladder network. A simplified block diagram of the processor is shown in Figure 10-25. An INTEL-MCS-80 design system was used, and the program for the test oscillator was stored in RAM via a TTY.

The test setup is shown in Figure 10-26 and includes the TTY as well as a small test fixture used to monitor the temperature. The test fixture also has the capability to force the digital output to any

---

*The processor itself was not included in the temperature chamber; however, the crystal oscillator as well as the A/D and D/A converters were exposed to the temperature change.

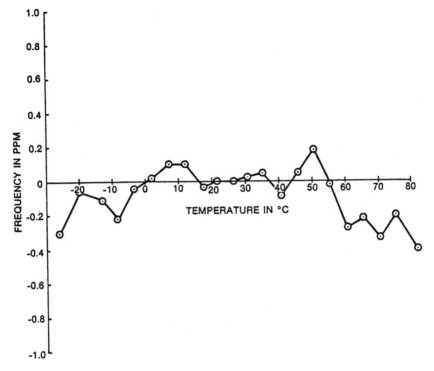

Figure 10-22. Frequency–Temperature for microprocessor compensated crystal oscillator (expanded scale).

value, thus allowing convenient data taking on the initial temperature run when the correction table is being determined. The oscillator is initially run over temperature and, at intervals of 5–10 degrees, the output switches are adjusted to put the oscillator on frequency. The temperature (address) is then read and recorded along with the correction required. The values used in this experiment are listed in hexadecimal notation, in Table 10-2. Photographs of the temperature chamber, the processor, test fixture, and oscillator are shown in Figures 10-27, 28, 29, and 30.

The microprocessor program was written in assembly language and requires 221 bytes of storage. In addition, 10 bytes of RAM are required as scratch-pad memory. A listing of the memory assignments is given in Tables 10-3 and 10-4.

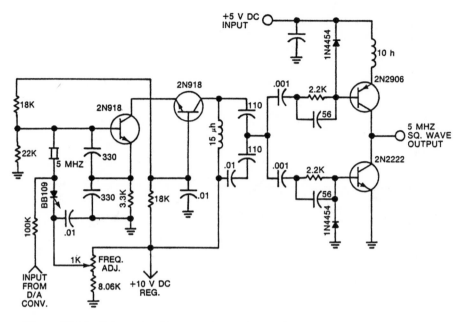

Figure 10-23. Schematic diagram of experimental oscillator using INTEL-8080 processor.

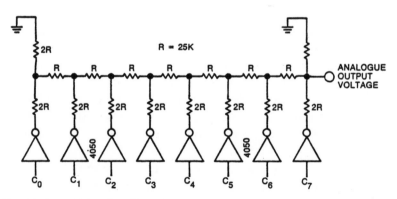

Figure 10-24. Digital-to-analog converter used with TCXO shown in Figure 10-23.

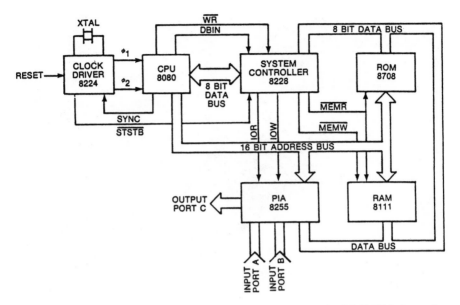

Figure 10-25. Simplified block diagram of processor used with TCXO shown in Figure 10-23.

Table 10-2. Stored Temperature–Frequency Correction Data.

Temp.	Corr.	Temp.	Corr.
00	00	78	A5
0A	61	82	8D
14	A3	8C	74
1E	CA	96	5D
28	DF	A0	45
32	ED	AA	30
3C	F0	B4	1E
46	EE	BE	12
50	E7	C8	0D
5A	DC	D2	0F
64	CC	D7	13
6E	BA	FF	FF

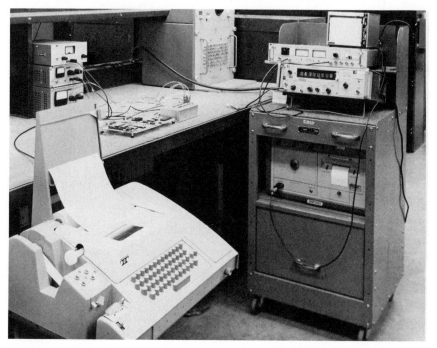

Figure 10-26. Test setup for microprocessor temperature compensation.

Figure 10-27. Temperature chamber used in experiment.

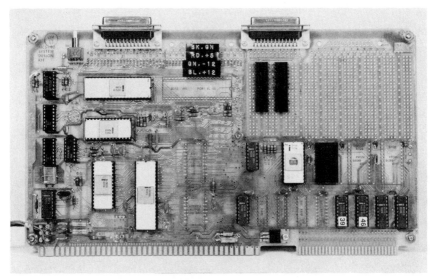

Figure 10-28. Processor used in experiment.

Figure 10-29. Close-up of test fixture and processor.

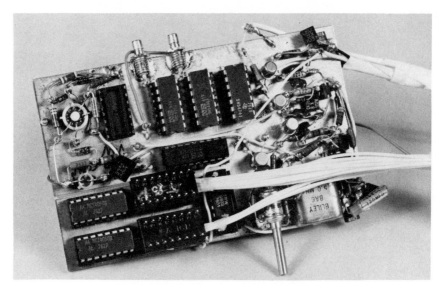

Figure 10-30. Oscillator used in experiment.

Table 10-3. Memory Usage.

Description	Location
Ambient temperature	
least significant byte (LSB)	PIAIA
most significant byte (MSB)	PIAIB
Stored correction data	
T1 (LSB)	DTA
T1 (MSB)	DTA+1
F1	DTA+2
T2 (LSB)	DTA+3
T2 (MSB)	DTA+4
F2 (LSB)	DTA+5
—	
—	
—	
—	
—	
End of file	
FF	DTA+n−1
FF	DTA+n
Correction voltage output	PIAOC

Table 10-4. Internal Memory and Register Usage.[a]

Description	Register/Memory
Ambient temperature	
T(LSB)	RAM
T(MSB)	RAM+1
Output to interpolation program	
TN(LSB)	RAM+3
TN(MSB)	RAM+4
FN	RAM+5
TN+1(LSB)	RAM+6
TN+1(MSB)	RAM+7
FN+1	RAM+8
FN+1−FN[b]	RAM+8
Working registers	
sign of $F_{n+1} - F_n$	RAM+2
dividend $T - T_n$	C
divisor $T_{n+1} - T_n$	D
quotient and multiplier	E
$(T - T_1)/(T_{n+1} - T_n)$	
multiplicand $(F_{n+1} - F_n)$	A
product	HL
counter for multiplication	B
and division	
store previous output	RAM+9

[a]A listing of the program is reproduced on the following pages.
[b]F_{n+1} in RAM+8 is destroyed by the program after $F_{n+1} - F_n$ is formed.

The program is loaded beginning at a location called ROM, which for the MCS-80 system was assigned a value 1200 in hexadecimal (H). The first scratch-pad location is called RAM and was assigned a value 1365H. The stored data table begins at a location called DTA and has a value 1300H. The 8255 peripheral interface adapter was wired so that port A, designated PIAIA, has an address 14H; port B, designated PIAIB, a value 15H; and port C, designated PIAOC, an address 16H.

The frequency accuracy of a crystal oscillator which is compensated using microprocessor techniques depends on many of the same factors as other methods of temperature compensation. Among these

```
8080 MACRO ASSEMBLER, VER 2.4              ERRORS = 0 PAGE 1

00001*   1365                    ORG 1365H
00002*   1365          RAM:      DS 9
00003*   1300                    ORG 1300H
00004*   1300          DTA:      DS 75
00005*   0017                    PTACR SET 17H
00006*   0014                    PTAIA SET 14H
00007*   0015                    PTAIB SET 15H
00008*   0016                    PTAOC SET 16H
00009*                           ;BEGINNING OF LINEAR SEARCH PROGRAM
00010*   1200                    ORG 1200H       ;ADDRESS OF FIRST INSTRUCTION IN PROGRAM
00011*   1200  3E92    ROM:      MVI A,92H  ;SET PIA FOR A AND B INPUT, C FOR OUTPUT
00012*   1202  D317              OUT PTACR  ; SET PIA A AND B FOR INPUT, C FOR OUTPUT
00013*   1204  DB14              IN PTAIA   ; READ IN A PORT OF PIA, LSB OF TEMP
00014*   1206  216513            LXI H,RAM
00015*   1209  77                MOV M, A
00016*   120A  DB15              IN PTAIB   ;READ IN B PORT OF PIA, MSB OF TEMP
00017*   120C  23                INX H
00018*   120D  77                MOV M,A
00019*   120E  216513            LXI H, RAM
00020*   1211  5E                MOV E, M
00021*   1212  23                INX H
00022*   1213  56                MOV D, M
00023*   1214  210013            LXI H,DTA       ;LOAD ADDRESS OF FIRST STORED COMPENSATION DATA
00024*   1217  7B      LOOP3:    MOV A,E         ;LOAD LSB OF ACTUAL TEMP INTO ACCUMULATOR
00025*   1218  96                SUB M           ;SUBTRACT T-TN LSB
00026*   1219  47                MOV B,A
00027*   121A  23                INX H           ;SET ADDRESS FOR MSB OF STORED TEMP
00028*   121B  7A                MOV A,D
00029*   121C  9E                SBB M           ;SUBTRACT MOST SIGNIFICANT BIT OF TEMP T-TN
00030*   121D  DA4012            JC X5           ;RESULT NEG READ OUT STORED POINTS
00031*   1220  CA2612            JZ X15
00032*   1223  C32C12            JMP X14
00033*   1226  78      X15:      MOV A, B
00034*   1227  C600              ADI 0
00035*   1229  CA3112            JZ X4
00036*   122C  23      X14:      INX H           ;SKIP FREQ CORRECTION
00037*   122D  23                INX H           ;INCREMENT ADDRESS TO NEXT STORED TEMP POINT
00038*   122E  C31712            JMP LOOP3       ;JUMP BACK TO BEGINNING OF SEARCH ROUTINE
00039*                           ;THE FOLLOWING INSTRUCTIONS WRITE OUT THE FREQ
00040*                           ;CORRECTION WORD IF THE SENSED TEMPERATURE
00041*                           ;COINSIDES WITH A STORED TEMPERATURE EXACTLY
00042*   1231  23      X4:       INX H           ;ADVANCE HL TO STORED FREQ CORRECTION
00043*   1232  7E                MOV A,M
00044*   1233  216E13            LXI H,RAM+9  ; LOAD ADDRESS OF PREVIOUS OUTPUT
00045*   1236  BE                CMP M        ; COMPARE WITH PREVIOUS OUTPUT
00046*   1237  CA0412            JZ ROM+4     ;SKIP PRINT OUT IF OUTPUT SAME AS BEFORE
00047*   123A  77                MOV M,A      ; STORE NEW OUTPUT IF DIFFERENT
00048*   123B  D316              OUT PTAOC
00049*   123D  C30412            JMP ROM+4    ;TRANSFER CONTROL TO BEGINNING OF PROGRAM
00050*   1240  2B      X5:       DCX H
```

are hystereses, temperature transients, the inaccuracy in determining temperature, and the inaccuracy in setting analog correction voltage to the desired value. The microprocessor has several advantages over other methods of compensation, however. Since a search for stored correction values can be made, it is reasonable to store points closer together over portions of the temperature range where the crystal has the largest slope. It is also possible to use various algorithms to calculate the correction required between stored points. In this discussion, a linear interpolation is assumed. It is possible, however, to develop algorithms based on more than the two closest stored values, such as fitting the cubic equation of the crystal.

```
00051*    1241   2B              DCX  H
00052*    1242   2B              DCX  H
00053*    1243   2B              DCX  H          ;SET HL TO ADDRESS OF CORRECTION DATA BELOW TEMP
00054*    1244   116813          LXI  D,RAM+3    ;LOAD ADDRESS WHERE DATA TO BE PLACED
00055*    1247   0606            MVI  B,6        ;SET B TO TRANSFER 6 BYTES OF DATA
00056*    1249   7E       X6:    MOV  A,M        ;READ DATA INTO ACCUMULATOR
00057*    124A   23              INX  H          ;INCREMENT SOURCE ADDRESS
00058*    124B   EB              XCHG            ;TRANSFER IN DESTINATION ADDRESS
00059*    124C   77              MOV  M,A        ;PLACE DATA AT DESTINATION ADDRESS
00060*    124D   23              INX  H          ;INCREMENT DESTINATION ADDRESS
00061*    124E   EB              XCHG            ;SET UP SOURCE ADDRESS
00062*    124F   05              DCR  B          ;DECREMENT COUNTER
00063*    1250   C24912          JNZ  X6         ;IF TRANSFER COMPLETE, CONINUE
00064*                                           ; BEGINNING OF INTERPOLATION PROGRAM
00065*    1253   216D13   BK1:   LXI  H,RAM+6    ;LOAD ADDRESS OF F2
00066*    1256   7E              MOV  A,M        ;LOAD F2 INTO ACCUMULATOR
00067*    1257   216A13          LXI  H,RAM+5    ;LOAD ADDRESS OF F1
00068*    125A   5E              MOV  E,M        ;PLACE F1 IN E
00069*    125B   BB              CMP  E          ;COMPARE A-E,I.E. SET CY IF F2 L.T. F1
00070*    125C   DA6712          JC   X7         ; JP IF F1 G.T. F2
00071*    125F   216713          LXI  H,RAM+2    ;LOAD ADDRESS OF FLAG
00072*    1262   3600            MVI  M,0        ;SIGN BIT 0 IF F1 L.T. F2
00073*    1264   C36F12          JMP  X8         ;SKIP INTERCHANGE IF  F1 L.T. F2
00074*    1267   57       X7:    MOV  D,A        ;INTERCHANGE E AND A
00075*    1268   7B              MOV  A,E
00076*    1269   5A              MOV  E,D
00077*    126A   216713          LXI  H,RAM+2    ;SET UP ADDRESS OF SIGN FLAG
00078*    126D   36FF            MVI  M,0FFH     ;SINCE F2-F1 IS NEG SET FLAG
00079*    126F   93       X8:    SUB  E          ;FORM ABS F2-F1
00080*    1270   216D13          LXI  H,RAM+6    ;SET UP ADDRESS TO STORE ABS F2-F1
00081*    1273   77              MOV  M,A        ;STORE DIFFERENCE
00082*    1274   216513          LXI  H,RAM      ;SET UP ADDRESS OF T LSB
00083*    1277   7E              MOV  A,M        ;ACCUMULATOR HOLDS LSB OF T
00084*    1278   216813          LXI  H,RAM+3    ;SET UP ADDRESS OF T1 LSB
00085*    127B   96              SUB  M          ;FORM T-T1
00086*    127C   4F       X16:   MOV  C,A
00087*    127D   216B13          LXI  H,RAM+6    ;SET UP ADDRESS OF T2 LSB
00088*    1280   7E              MOV  A,M        ;T2 LSB IN ACC
00089*    1281   216813          LXI  H,RAM+3    ;SET UP ADDRESS OF T1 LSB
00090*    1284   96              SUB  M          ;ACCUMULATOR HOLDS T2-T1
00091*    1285   57       X17:   MOV  D,A
00092*                                           ;THE FOLLOWING INSTRUCTIONS CONTAIN
00093*                                           ;THE 8 BIT DIVISION PROGRAM
00094*    1286   1E00     BK2:   MVI  E,0        ;INITIALIZE QUITENT TO ZERO
00095*    1288   0609            MVI  B,9H       ;SET UP COUNTER FOR 8 SHIFTS
00096*    128A   79       X11:   MOV  A,C        ;PLACE DIVIDEND IN ACC LSB
00097*    128B   92              SUB  D          ;TRY SUBTRACTION
00098*    128C   DA9812          JC   X9         ;JUMP IF SUBTRACTION UNSUCCESSFUL
00099*    128F   17              RAL             ;SHIFT DIVIDEND
00100*    1290   4F              MOV  C,A        ;MOVE SHIFTED DIVIDEND BACK TO C
00101*    1291   7B              MOV  A,E
```

For this discussion, let the frequency error at any given temperature be represented by

$$E = St_{\text{lsb}} + \frac{F}{2^b} + \delta \text{ ppm} \qquad (10\text{-}30)$$

where

E is the overall frequency error in parts per million;

S is the maximum slope of the frequency–temperature curve of the oscillator in parts per million per Celsius degree;

t_{lsb} is the temperature range represented by the least significant bit of the digital temperature input;

F is the total frequency pulling range of the varactor;

```
00102*    1292    37              STC         ISET CARRY
00103*    1293    17              RAL         IROTATE CARRY INTO QUOTENT
00104*    1294    5F              MOV E,A     IMOVE QUOTENT BACK TO E
00105*    1295    C3A012          JMP X10
00106*    1298    79       X9:    MOV A,C     IMOV C INTO ACC TO RESTORE
00107*    1299    3F              CMC         ICOMPLEMENT CARRY I.E. SET CY=0
00108*    129A    17              RAL         I ROTATE DIVIDEND
00109*    129B    4F              MOV C,A     IRESTORE LSB OF DIVIDEND
00110*    129C    AF              XRA A       ICLEAR CARRY
00111*    129D    7B              MOV A,E     IPLACE QUOTENT IN ACC
00112*    129E    17              RAL         IROTATE ZERO INTO QUQTENT
00113*    129F    5F              MOV E,A     IREPLACE ROTATED QUOTENT IN E
00114*    12A0    05       X10:   DCR B       IDECREMENT COUNTER
00115*    12A1    C2AA12          JNZ X11     IPROCEED WITH NEW SUBTRACTION
00116*                                        ITHE FOLLOWING INSTRUCTIONS PERFORM A 16X8
00117*                                        IBIT MULTIPLICATION
00118*    12A4    1600     BK3:   MVI D,0     ISET MSB MULTIPLICAND TO ZERO
00119*    12A6    216D13          LXT H,RAM+8 ISET UP ADDRESS OF F2-F1
00120*    12A9    7E              MOV A,M     IMOV F2-F1 INTO ACCUMULATOR
00121*    12AA    210000          LXI H,0     IINITIALIZE PRODUCT TO ZERO
00122*    12AD    0608            MVI B,8     ISET UP CONTROL LOOP FOR 8 OPERATIONS
00123*    12AF    29       LOOP:  DAD H       ISHIFT PARTIAL PRODUCT LEFT AND INTO CARRY
00124*    12B0    17              RAL         IROTATE MULTIPLIER BIT TO CARRY
00125*    12B1    D2B712          JNC DEC     ITEST MULTIPLIER AT CARRY
00126*    12B4    19              DAD D       IADD MULTIPLICAND TO PARTIAL PRODUCT IF CY=1
00127*    12B5    CE00            ACI 0       IADD CARRY
00128*    12B7    05       DEC:   DCR B       IDECREMENT B LOOP COUNTER
00129*    12B8    C2AF12          JNZ LOOP    IREPEAT IF NOT 8 TIMES
00130*    12BB    EB              XCHG        IPLACE PRODUCT IN DE
00131*    12BC    216713          LXI H, RAM+2 ILOAD ADDRESS OF SIGN
00132*    12BF    AF              XRA A       ICLEAR A
00133*    12C0    86              ADD M       IBRING SIGN BIT INTO ACC
00134*    12C1    FACC12          JM X13      IJUMP IF NEGATIVE PRODUCT
00135*    12C4    216A13          LXI H,RAM+5 ILOAD ADDRESS OF F1
00136*    12C7    7E              MOV A,M     IF1 TO ACCUMULATOR
00137*    12C8    82              ADD D       IFORM CORRECTION
00138*    12C9    C3D112          JMP X12     IGO TO READ OUT INSTRUCTIONS
00139*    12CC    216A13   X13:   LXI H,RAM+5 ILOAD ADDRESS OF F1
00140*    12CF    7E              MOV A,M     IF1 IN ACC
00141*    12D0    92              SUB D       IFORM CORRECTION
00142*    12D1    216E13   X12:   LXI H,RAM+9 I LOAD ADDRESS OF PREVIOUS OUTPUT
00143*    12D4    BE              CMP M       I COMPARE WITH PREVIOUS OUTPUT
00144*    12D5    CA0412          JZ ROM+4    ISKIP PRINT OUT IF OUTPUT SAME AS BEFORE
00145*    12D8    77              MOV M,A     I STORE NEW OUTPUT IF DIFFERENT
00146*    12D9    D316            OUT PIAOC
00147*    12DB    C30412   BK4:   JMP ROM+4   ISTART OVER
00148*                           END
NO PROGRAM ERRORS
```

b is the number of bits in the digital correction word; and

δ is the error resulting from the spacing between stored correction values.

If a linear interpolation program is used, the error δ is the difference between the actual crystal curve and a straight line joining the two nearest stored correction values.

It is assumed here that a linear modulator is used so that the correction voltage is proportional to the frequency correction required.

It is well known that the frequency–temperature curve for a quartz crystal is a cubic equation of the form

$$\frac{\Delta f}{f} = A_1 X + A_2 X^2 + A_3 X^3 \text{ ppu} \qquad (10\text{-}31)$$

8080 MACRO ASSEMBLER, VER 2.4   FRRORS = 0 PAGE 4

                              SYMPOL TARLE

    • 01

    A   0007   B   0000   HK1   1253 •   RK2   1286 •
    BK3   12A4 • BK4   1208 • C   0001   D   0002
    DEC   12B7   D1A   1300   F   0003   H   0004
    L   0005   LOOP   12AF   LOOPX   1217   M   0006
    FIACH   0017   P1AIA   0014   PIAIB   0015   PIAOC   0016
    PSW   0006   RAM   1365   ROM   1200   SP   0006
    X10   12A0   X11   128A   X12   1201   X13   12CC
    X14   122C   X15   1226   X16   127C •   X17   1285 •
    X4   1231   X5   1240   X6   1249   X7   1267
    X8   126F   X9   1298

where $X = t - t_0$. The coefficients vary slightly with the crystal parameters and a comprehensive listing is given by Bechmann in ref. 4. From this reference one set of values for plated resonators using natural quartz are given below. For a fundamental-mode AT-cut crystal,

$$A_1 \doteq (\text{ANG}) (-5.15 \times 10^{-6}) \tag{10-32}$$

$$A_2 \doteq (\text{ANG}) (-4.5 \times 10^{-9}) - 0.1 \times 10^{-9} \tag{10-33}$$

$$A_3 \doteq (\text{ANG}) (-20 \times 10^{-12}) + 130 \times 10^{-12} \tag{10-34}$$

where ANG is the angle in degrees of arc from the angle of cut to give a zero slope at t_0. (The term t is the temperature in Celsius degrees and t_0 is taken to be 20°C.)

For a third overtone AT-cut crystal,

$$A_1 \doteq (\text{ANG}) (-5.15 \times 10^{-6}) \tag{10-35}$$

$$A_2 \doteq (\text{ANG}) (-4.5 \times 10^{-9}) - 1.7 \times 10^{-9} \tag{10-36}$$

$$A_3 \doteq 105 \times 10^{-12} \tag{10-37}$$

and for a fifth overtone,

$$A_1 \doteq (\text{ANG}) (-5.5 \times 10^{-6}) \tag{10-38}$$

$$A_2 \doteq (\text{ANG}) (-4.5 \times 10^{-9}) - 1.2 \times 10^{-9} \tag{10-39}$$

$$A_3 \doteq 105 \times 10^{-12} \tag{10-40}$$

If we substitute $X = t - 20$ into equation (10-31) we obtain

$$\frac{\Delta f(t)}{f} = b_1 t + b_2 t^2 + b_3 t^3 + b_4 \tag{10-41}$$

where
$$b_1 = A_1 - 40A_2 + 1200A_3 \tag{10-42}$$
$$b_2 = A_2 - 60A_3 \tag{10-43}$$
$$b_3 = A_3 \tag{10-44}$$
$$b_4 = 400A_2 - 20A_1 - 8000A_3. \tag{10-45}$$

The microprocessor uses a linear interpolation algorithm of the form

$$f(t_1) + \frac{f(t_2) - f(t_1)}{(t_2 - t_1)}(t - t_1), \tag{10-46}$$

and the frequency error over the interval t_1 to t_2 is given by the difference between equations (10-41) and (10-46) and is

$$\delta = f(t) - f(t_1) - \frac{[f(t_2) - f(t_1)](t - t_1)}{(t_2 - t_1)}. \tag{10-47}$$

It is convenient in dealing with this equation to perform a coordinate transformation, as shown in Figure 10-31. Using the dotted axes we have

$$F(T) = f(t) - f(t_1) \quad \text{and} \quad T = t - t_1.$$

The equation (10-41) becomes

$$F(T) = b_1(T + t_1) + b_2(T + t_1)^2 + b_3(T + t_1)^3 + b_4 - f(t_1). \tag{10-48}$$

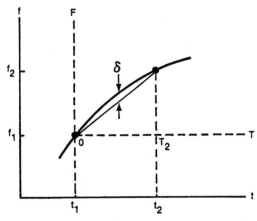

Figure 10-31. Frequency–Temperature curve transformation.

Expanding and simplifying gives

$$F(T) = (b_1 + 2b_2 t_1 + 3b_3 t_1^2) T + (b_2 + 3b_3 t_1) T^2 + b_3 T^3. \qquad (10\text{-}49)$$

We may then write equation (10-47) as

$$\delta = F(T) - \frac{F(T_2) T}{T_2}, \qquad 0 \leqslant T \leqslant T_2. \qquad (10\text{-}50)$$

Substituting from equation (10-49) we have

$$\delta = (b_1 + 2b_2 t_1 + 3b_3 t_1^2) T + (b_2 + 3b_3 t_1) T^2 + b_3 T^3 - [(b_1 + 2b_2 t_1$$
$$+ 3b_3 t_1^2) T_2 + (b_2 + 3b_3 t_1) T_2^2 + b_3 T_2^3] \frac{T}{T_2}, \qquad (10\text{-}51)$$

which can be simplified to

$$\delta = -(b_2 T_2 + 3b_3 t_1 T_2 + b_3 T_2^2) T + (b_2 + 3b_3 t_1) T^2 + b_3 T^3. \qquad (10\text{-}52)$$

Obviously, $\delta = 0$ at $T = 0$ and $T = T_2$ because of the way the expression for δ was constructed. A maximum or minimum occurs near the center of the interval and is found by setting $d\delta/dT = 0$.
Let

$$K_1 = -(b_2 T_2 + 3b_3 t_1 T_2 + b_3 T_2^2) \qquad (10\text{-}53)$$

$$K_2 = (b_2 + 3b_3 t_1). \qquad (10\text{-}54)$$

Then

$$\delta = K_1 T + K_2 T^2 + b_3 T^3 \qquad (10\text{-}55)$$

$$\frac{d\delta}{dT} = K_1 + 2K_2 T + 3b_3 T^2. \qquad (10\text{-}56)$$

Setting this expression to zero and solving for T gives

$$T = -\frac{2K_2 \pm \sqrt{4K_2^2 - 12b_3 K_1}}{6b_3} \qquad (10\text{-}57)$$

$$T = \frac{-K_2 [1 \mp \sqrt{1 - (3b_3 K_1/K_2^2)}]}{3b_3}. \qquad (10\text{-}58)$$

However, if the intervals are small so that

$$\left| \frac{3K_1 b_3}{K_2^2} \right| = \left| \frac{-3b_3 (b_2 T_2 + 3b_3 t_1 T_2 + b_3 T_2^2)}{(b_2 + 3b_3 t_1)^2} \right| \ll 1 \quad (10\text{-}59)$$

we may expand $\sqrt{1 - (3b_3 K_1/K_2^2)}$ using the binomial series and retain only the first two terms; thus

$$(1 + m)^\alpha \doteq 1 + \alpha m \quad \text{for} \quad m \ll 1$$

and equation (10-58) becomes

$$T = \frac{-K_2}{3b_3}\left(1 - 1 + \frac{3b_3 K_1}{2K_2^2}\right) \tag{10-60}$$

$$T = -\frac{K_1}{2K_2}. \tag{10-61}$$

Now substituting from (10-53) and (10-54)

$$T = \frac{b_2 T_2 + 3b_3 t_1 T_2 + b_3 T_2^2}{2(b_2 + 3b_3 t_1)}. \tag{10-62}$$

If the temperature interval is small as previously assumed,

$$(b_2 + 3b_3 t_1) \gg b_3 T_2 \quad \text{and} \quad T \doteq \frac{T_2(b_2 + 3b_3 t_1)}{2(b_2 + 3b_3 t_1)} = \frac{T_2}{2}. \tag{10-63}$$

Thus the maximum error occurs in the center of the temperature interval. Substituting this into equation (10-52) gives:

$$\delta_{max} = -(b_2 T_2 + 3b_3 t_1 T_2 + b_3 T_2^2)\frac{T_2}{2} + (b_2 + 3b_3 t_1)\frac{T_2^2}{4} + \frac{b_3 T_2^3}{8}, \tag{10-64}$$

which may be simplified to

$$\delta_{max} = \frac{-T_2^2}{4}\left[\frac{3}{2}b_3 T_2 + b_2 + 3b_3 t_1\right]. \tag{10-65}$$

Values of δ_{max} are given in Table 10-5 for temperature intervals of 3°C, 6°C, and 10°C using a typical TCXO crystal with ANG = 7 minutes of arc (0.1167 deg). Then from equations (10-32) to (10-34) and (10-42) to (10-45) we have

$$b_1 = -4.228 \times 10^{-7}$$

$$b_2 = -8.287 \times 10^{-9}$$

Table 10-5. Maximum Tracking Error for Stored
Temperature Intervals of 3°C, 6°C, and 10°C.

Ambient temp. t_1 (°C)	δ_{max} (ppm)		
	$T_2 = 3°C$	$T_2 = 6°C$	$T_2 = 10°C$
-55	0.0648	0.254	0.686
-30	0.0432	0.167	0.447
-15	0.0303	0.116	0.303
0	0.0174	0.0642	0.159
10	0.0087	0.0298	0.0635
15	0.0044	0.0125	0.0157
25	-0.0042	-0.0219	-0.0800
50	-0.0257	-0.108	-0.319
65	-0.0387	-0.167	-0.463
80	-0.0516	-0.212	-0.607
95	-0.0645	-0.263	-0.750

$$b_3 = 1.2766 \times 10^{-10}$$

$$b_4 = 1.074 \times 10^{-5}.$$

It is possible, of course, to use the microprocessor to calculate the frequency correction directly from the cubic equation of the crystal or to use other algorithms taking advantage of more than the two nearest stored correction values. These methods may be used in connection with a fine-correction lookup table to compensate for minor irregularities in the crystal curve.

Temperature compensation using microprocessors is subject to the same ultimate limitations on accuracy as the digital method of compensation, namely temperature transients and frequency hysteresis. At the time of this writing the limit is about ±5 parts in 10^8 for an AT-cut crystal, because of hysteresis. As better crystals are developed it will be possible to make additional improvements in the compensation accuracy by the addition of more bits in the temperature and frequency correction words.

Research to improve the accuracy of TCXOs has been carried out in various laboratories for over two decades and will no doubt continue for some years to come. The requirement for highly accurate frequency and time standards which are also low-cost, low-power, and small in size is considerable, and promises many benefits to a broad spectrum of the electronics industry, including both military and commercial areas as well as in data processing and transmission equipment. Many techniques remain to be tried, and the use of microprocessors will no doubt play an important roll in implementing many of them.

Appendix A
Derivation of the Complex Equation for Oscillation

Using the block diagram of Figure A-1, in which the active element is represented by its Y-parameters and the feedback network by its Z-parameters, the complex equation for oscillation can be derived. By the definition of Y-parameters, the currents and voltages of the active device can be described by the following equations:

$$I = y_{11} V + y_{12} V' \tag{A-1}$$

$$I' = y_{21} V + y_{22} V'. \tag{A-2}$$

Also by definition, the currents and voltages of the feedback network can be described by the equations:

$$V' = -Z_{11} I' - Z_{12} I \tag{A-3}$$

$$V = -Z_{21} I' - Z_{22} I. \tag{A-4}$$

Arranging the equations symmetrically,

$$V y_{11} + V' y_{12} - I + 0 I' = 0 \tag{A-5}$$

$$V y_{21} + V' y_{22} + 0 I - I' = 0 \tag{A-6}$$

$$0 V + V' + I Z_{12} + I' Z_{11} = 0 \tag{A-7}$$

$$V + 0 V' + I Z_{22} + I' Z_{21} = 0. \tag{A-8}$$

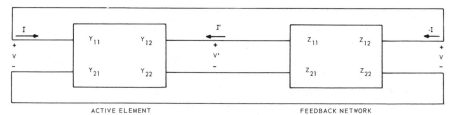

ACTIVE ELEMENT   FEEDBACK NETWORK

Figure A-1. General oscillator: block diagram.

Solving these equations for V using determinants,

$$V = \frac{\begin{vmatrix} 0 & y_{12} & -1 & 0 \\ 0 & y_{22} & 0 & -1 \\ 0 & 1 & Z_{12} & Z_{11} \\ 0 & 0 & Z_{22} & Z_{21} \end{vmatrix}}{\begin{vmatrix} y_{11} & y_{12} & -1 & 0 \\ y_{21} & y_{22} & 0 & -1 \\ 0 & 1 & Z_{12} & Z_{11} \\ 1 & 0 & Z_{22} & Z_{21} \end{vmatrix}}. \tag{A-9}$$

It will be seen that the numerator of this expression is zero, making $V = 0$ for every case except that for which the denominator also is zero; in that case, V is indeterminate. We know, however, that if oscillation takes place, $V \neq 0$, and therefore it must be true that

$$\begin{vmatrix} y_{11} & y_{12} & -1 & 0 \\ y_{21} & y_{22} & 0 & -1 \\ 0 & 1 & Z_{12} & Z_{11} \\ 1 & 0 & Z_{22} & Z_{21} \end{vmatrix} = 0. \tag{A-10}$$

Solving this determinant gives the equation:

$$y_{21}Z_{21} + y_{11}Z_{22} + y_{22}Z_{11} + y_{12}Z_{12} + \Delta y \Delta Z + 1 = 0 \tag{A-11}$$

where

$$\Delta y = y_{11}y_{22} - y_{21}y_{12}$$

$$\Delta Z = Z_{11}Z_{22} - Z_{21}Z_{12}.$$

This equation is quite general and may be applied to almost any oscillator. If we choose, we may represent the active device by Y-parameters and the feedback network by Z-parameters or vice versa. It is usually more convenient to use the former, however.

It should be pointed out that the use of the two-port parameters implies that the circuits are linear. At large-signal amplitudes, the Y-parameters therefore must be defined as the ratios of fundamental components of current to fundamental components of voltage.

In working with transistors, it may be convenient to use the convention $y_{11} = y_i$, $y_{21} = y_f$, $y_{12} = y_r$, and $y_{22} = y_o$, which conforms to present usage on transistor data sheets. Making the same transition in the Z-parameter gives equation

(A-11) as

$$y_f Z_f + y_i Z_o + y_o Z_i + y_r Z_r + \Delta y \Delta Z + 1 = 0, \qquad \text{(A-12)}$$

where

$$\Delta y = y_i y_o - y_f y_r$$
$$\Delta Z = Z_i Z_o - Z_f Z_r,$$

which is the form used throughout this book. [Equation (A-12) is presented and discussed at some length in reference 41.]

Appendix B
Derivation of *Y*-Parameter Equations for the Pierce Oscillator

The Pierce oscillator can be represented by the schematic diagram of Figure B-1.

Let the circuit be redrawn and the transistor replaced by a four-terminal network described by its *Y*-parameters. The diagram of Figure B-2 then results.

If the transistor is represented by its *Y*-parameters, then the feedback network must be represented by its *Z*-parameters if equation (A-12) is to be used. For this analysis, it is convenient to include the load resistance R_L in the output admittance of the transistor. This is accomplished by using a value of y_{oe} which is equal to y_{oe} (transistor) $+ 1/R_L$. Also, the input capacity of the transistor will be lumped in parallel with C_1, thus making $y_{ie} = g_{ie}$ purely resistive in the analysis. In like manner, the output capacity of the transistor will be

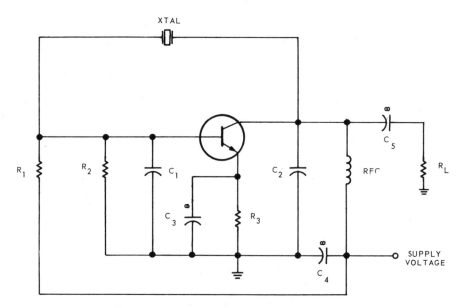

Figure B-1. Pierce oscillator: schematic diagram.

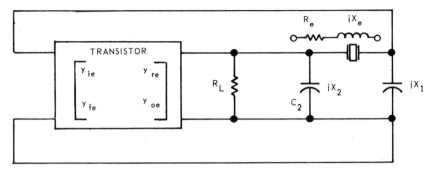

Figure B-2. Pierce oscillator: ac circuit diagram.

lumped in parallel with C_2, making $y_{oe} = g_{oe}$ purely resistive. In the analysis it will be assumed that the reverse transfer admittance of the transistor is purely imaginary: $y_{re} = jb_{re}$.

The remaining π-network is shown in Figure B-3.

From this circuit the Z-parameters can be calculated using the following definitions:

Input impedance $\qquad\qquad Z_i = \dfrac{V_1}{I_1}\bigg|_{I_2=0}$ $\qquad\qquad$ (B-1)

Output impedance $\qquad\qquad Z_o = \dfrac{V_2}{I_2}\bigg|_{I_1=0}$ $\qquad\qquad$ (B-2)

Forward transfer impedance $\qquad\qquad Z_f = \dfrac{V_2}{I_1}\bigg|_{I_2=0}$ $\qquad\qquad$ (B-3)

Reverse transfer impedance $\qquad\qquad Z_r = \dfrac{V_1}{I_2}\bigg|_{I_1=0}$ $\qquad\qquad$ (B-4)

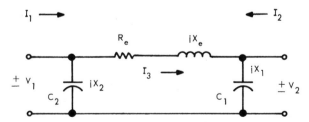

Figure B-3. Pierce oscillator: π-network: simplified diagram.

From equation (B-1) it can be seen that Z_i is merely the input impedance of the network with the output open-circuited and can be written by inspection as

$$Z_i = \frac{jx_2(R_e + jx_e + jx_1)}{jx_2 + (R_e + jx_e + jx_1)}. \qquad \text{(B-5)}$$

If we let

$$Z = R_e + j(x_1 + x_2 + x_e) \qquad \text{(B-6)}$$

we have

$$Z_i = \frac{1}{Z} jx_2(R_e + jx_e + jx_1) \qquad \text{(B-7)}$$

$$Z_i = -\frac{1}{Z}(x_1 x_2 + x_2 x_e - jR_e x_2). \qquad \text{(B-8)}$$

By symmetry, we may write:

$$Z_o = -\frac{1}{Z}[x_1 x_2 + x_1 x_e - jR_e x_1]. \qquad \text{(B-9)}$$

The forward transfer impedance is given by the ratio of V_2 to I_1 with the output open-circuited. In order to calculate this, it is convenient to first determine I_3 in terms of I_1, which is given by

$$I_3 = \frac{I_1(jx_2)}{(R_e + jx_e + jx_1) + jx_2} \qquad \text{(B-10)}$$

$$I_3 = \frac{I_1(jx_2)}{Z} \qquad \text{(B-11)}$$

and

$$V_2 = I_3(jx_1); \qquad \text{(B-12)}$$

therefore,

$$V_2 = \frac{I_1(jx_2)}{Z} jx_1 \qquad \text{(B-13)}$$

and

$$\frac{V_2}{I_1} = \frac{-x_1 x_2}{Z} = Z_f. \qquad \text{(B-14)}$$

Since the π-network is composed entirely of reciprocal elements, the forward and reverse transfer impedances are equal, so that

$$Z_f = Z_r = \frac{-x_1 x_2}{Z}. \tag{B-15}$$

In using equation (A-12) the quantity ΔZ must be used. It is given by

$$\Delta Z = Z_i Z_o - Z_f Z_r \tag{B-16}$$

$$\Delta Z = -\frac{1}{Z}(x_1 x_2 + x_2 x_e - jR_e x_2)\left(-\frac{1}{Z}\right)(x_1 x_2 + x_1 x_e - jR_e x_1)$$

$$- \left(-\frac{x_1 x_2}{Z}\right)\left(-\frac{x_1 x_2}{Z}\right) \tag{B-17}$$

$$\Delta Z = \frac{1}{Z^2}\left[(x_1 x_2 + x_2 x_e)(x_1 x_2 + x_1 x_e) - j(x_1 x_2 + x_2 x_e)R_e x_1\right.$$

$$\left. - j(x_1 x_2 + x_1 x_e)R_e x_2 - R_e^2 x_1 x_2\right] - \frac{x_1^2 x_2^2}{Z^2} \tag{B-18}$$

$$\Delta Z = \frac{1}{Z^2}(x_1^2 x_2^2 + x_1^2 x_2 x_e + x_2^2 x_1 x_e + x_1 x_2 x_e^2 - jR_e x_1^2 x_2 - jR_e x_1 x_2 x_e$$

$$- jR_e x_1 x_2^2 - jR_e x_1 x_2 x_e - R_e^2 x_1 x_2 - x_1^2 x_2^2) \tag{B-19}$$

$$\Delta Z = \frac{1}{Z^2}\{R_e x_1 x_2 [-R_e - j(x_2 + x_e + x_1)] + jx_1 x_2 x_e [-R_e - jx_1 - jx_2 - jx_e]\} \tag{B-20}$$

$$\Delta Z = -\frac{1}{Z}(R_e x_1 x_2 + jx_1 x_2 x_e) \tag{B-21}$$

$$\Delta Z = -\frac{x_1 x_2}{Z}(R_e + jx_e). \tag{B-22}$$

Summarizing these results, we have

$$Z_i = -\frac{1}{Z}(x_1 x_2 + x_2 x_e - jR_e x_2) \tag{B-23}$$

$$Z_f = Z_r = -\frac{x_1 x_2}{Z} \tag{B-24}$$

$$Z_o = -\frac{1}{Z}(x_1 x_2 + x_1 x_e - jR_e x_1) \tag{B-25}$$

$$\Delta Z = -\frac{x_1 x_2}{Z}(R_e + jx_e), \tag{B-26}$$

where

$$Z = R_e + j(x_1 + x_2 + x_e). \tag{B-27}$$

These results can be substituted in the general equation for oscillation as developed in appendix A.

$$y_{fe}Z_f + y_{ie}Z_o + y_{oe}Z_i + y_{re}Z_r + \Delta y\Delta Z + 1 = 0 \tag{B-28}$$

$$-y_{fe}\frac{(x_1 x_2)}{Z} - y_{ie}\frac{(x_1 x_2 + x_1 x_e - jR_e x_1)}{Z} - y_{oe}\frac{(x_1 x_2 + x_2 x_e - jR_e x_2)}{Z}$$

$$-y_{re}\frac{(x_1 x_2)}{Z} - \Delta y\frac{(x_1 x_2)(R_e + jx_e)}{Z} + 1 = 0. \tag{B-29}$$

Multiplying by $-Z$ gives

$$y_{fe}x_1 x_2 + y_{ie}(x_1 x_2 + x_1 x_e - jR_e x_1) + y_{oe}(x_1 x_2 + x_2 x_e - jR_e x_2)$$

$$+ y_{re}x_1 x_2 + \Delta y x_1 x_2 (R_e + jx_e) - Z = 0. \tag{B-30}$$

Making the substitutions:

$$y_{fe} = g_{fe} + jb_{fe}, \quad y_{ie} = g_{ie}, \quad y_{oe} = g_{oe}, \quad y_{re} = jb_{re},$$

and

$$\Delta y = (y_{ie}y_{oe} - y_{fe}y_{re}) = g_{ie}g_{oe} - jb_{re}(g_{fe} + jb_{fe}),$$

and for Z results in the equation

$$(g_{fe} + jb_{fe})x_1 x_2 + g_{ie}(x_1 x_2 + x_1 x_e - jR_e x_1)$$

$$+ g_{oe}(x_1 x_2 + x_2 x_e - jR_e x_2) + jb_{re}x_1 x_2$$

$$+ [g_{ie}g_{oe} - jb_{re}(g_{fe} + jb_{fe})] x_1 x_2 (R_e + jx_e)$$

$$-R_e - j(x_1 + x_2 + x_e) = 0, \tag{B-31}$$

or

$$g_{fe}x_1 x_2 + jb_{fe}x_1 x_2 + g_{ie}x_1 x_2 + g_{ie}x_1 x_e - jg_{ie}R_e x_1$$

$$+ g_{oe}x_1 x_2 + g_{oe}x_2 x_e - jg_{oe}R_e x_2 + jb_{re}x_1 x_2$$

$$+ g_{ie}g_{oe}x_1 x_2 R_e + jg_{ie}g_{oe}x_1 x_2 x_e - jb_{re}g_{fe}x_1 x_2 R_e$$

$$+ b_{re}g_{fe}x_1x_2x_e + b_{re}b_{fe}x_1x_2R_e + jb_{re}b_{fe}x_1x_2x_e$$
$$- R_e - j(x_1 + x_2 + x_e) = 0. \tag{B-32}$$

The equation can be separated into real and imaginary components. The real parts of the equation are

$$g_{fe}x_1x_2 + g_{ie}x_1(x_2 + x_e) + g_{oe}x_2(x_1 + x_e) + R_e x_1 x_2(g_{ie}g_{oe} + b_{fe}b_{re})$$
$$+ g_{fe}b_{re}x_1x_2x_e - R_e = 0. \tag{B-33}$$

The imaginary parts are

$$-x_1 - x_2 - x_e + (b_{fe} + b_{re})x_1x_2 + g_{ie}g_{oe}x_1x_2x_e - R_e(g_{ie}x_1 + g_{oe}x_2)$$
$$+ b_{re}b_{fe}x_1x_2x_e - g_{fe}b_{re}x_1x_2R_e = 0. \tag{B-34}$$

Rewriting these equations and separating the primary and secondary effects gives

$$g_{fe}x_1x_2 = R_e + K_1 \tag{B-35}$$

and

$$x_1 + x_2 + x_e = 0 + K_2, \tag{B-36}$$

where

$$K_1 = -g_{ie}x_1(x_2 + x_e) - g_{oe}x_2(x_1 + x_e)$$
$$-R_e x_1 x_2(g_{ie}g_{oe} + b_{fe}b_{re})$$
$$- g_{fe}b_{re}x_1x_2x_e,$$

and

$$K_2 = (b_{fe} + b_{re})x_1x_2 + g_{ie}g_{oe}x_1x_2x_e$$
$$- R_e(g_{ie}x_1 + g_{oe}x_2) + b_{re}b_{fe}x_1x_2x_e$$
$$- g_{fe}b_{re}x_1x_2R_e.$$

If we assume that K_2 is zero, then

$$x_e = -(x_1 + x_2) \tag{B-37}$$

or

$$x_e = \frac{1}{\omega C_1} + \frac{1}{\omega C_2}. \tag{B-38}$$

Let T be some parameter which causes C_1 to vary with a rate $\partial C_1/\partial T$, and C_2 with a rate $\partial C_2/\partial T$. It is assumed here that C_1 and C_2 include the transistor input and output capacitances which are primarily responsible for the changes in C_1 and C_2. Then

$$\frac{\partial C_1}{\partial T} \doteq \frac{\partial C_{in}}{\partial T} \quad \text{and} \quad \frac{\partial C_2}{\partial T} \doteq \frac{\partial C_{out}}{\partial T}.$$

Differentiating equation (B-38) gives

$$\frac{\partial x_e}{\partial T} = -\frac{1}{\omega C_1^2}\left(\frac{\partial C_1}{\partial T}\right) - \frac{1}{\omega C_2^2}\left(\frac{\partial C_2}{\partial T}\right). \tag{B-39}$$

Also, from equation (B-35), if we assume that K_1 is zero, $C_1 = g_{fe}/R_e C_2 \omega^2$. Putting this in equation B-39 gives

$$\frac{\partial x_e}{\partial T} = -\left[\frac{R_e^2 C_2^2 \omega^3}{g_{fe}^2}\left(\frac{\partial C_1}{\partial T}\right)\right] - \left[\frac{1}{\omega C_2^2}\left(\frac{\partial C_2}{\partial T}\right)\right]. \tag{B-40}$$

If this expression is minimized for C_2, then

$$\frac{\partial(\partial x_e/\partial T)}{\partial C_2} = -\frac{2R_e^2 C_2 \omega^3}{g_{fe}^2}\left(\frac{\partial C_1}{\partial T}\right) + \frac{2}{\omega C_2^3}\left(\frac{\partial C_2}{\partial T}\right) = 0. \tag{B-41}$$

Solving this for C_2 gives

$$C_2^4 = \frac{(\partial C_2/\partial T)}{(\partial C_1/\partial T)} \times \frac{g f_e^2}{\omega^4 R_e^2}. \tag{B-42}$$

In a similar manner, it can be shown that

$$C_1^4 = \frac{(\partial C_1/\partial T)}{(\partial C_2/\partial T)} \times \frac{g f_e^2}{\omega^4 R_e^2}. \tag{B-43}$$

Dividing equation (B-42) by (B-43) gives

$$\left(\frac{C_2}{C_1}\right)^4 = \frac{(\partial C_2/\partial T)^2}{(\partial C_1/\partial T)^2} \tag{B-44}$$

or

$$\frac{C_2}{C_1} = \left(\frac{\partial C_2/\partial T}{\partial C_1/\partial T}\right)^{1/2}. \tag{B-45}$$

This minimizes the frequency change with respect to some parameter T provided the assumption $K_1 = K_2 = 0$ is valid.

A close examination of equation (B-41) shows that the condition

$$\frac{\partial(\partial x_e/\partial T)}{\partial C_2} = 0$$

does not necessarily assure that a minimum occurs. It does, however, assure that a minimum value of $|\partial x_e/\partial T|$ occurs when $\partial C_1/\partial T$ and $\partial C_2/\partial T$ are of like sign. If they are opposite sign $\partial(\partial x_e/\partial T)/\partial C_2 \neq 0$, the quantity $\partial x_e/\partial T$ then may be set equal to zero, and solving equation (B-39) for this condition gives the result

$$\frac{C_2}{C_1} = \left(-\frac{\partial C_2/\partial T}{\partial C_1/\partial T}\right)^{1/2}. \qquad (B-46)$$

Appendix C

Derivation of Y-Parameter Equations for the Grounded-Base Oscillator

The grounded-base crystal oscillator can be represented by the schematic diagram of Figure C-1.

Let the circuit be redrawn and the transistor replaced by a four-terminal network described by its Y-parameters. The diagram of Figure C-2 then results. If the transistor is represented by its Y-parameters, the feedback network must be represented by its Z-parameters if equation (A-12) is to be used.

For this analysis it is convenient to include the output resistance of the transistor in the load resistor R_T. This is accomplished by using a value of R_T which is equal to $R_L + (1/g_{ob})$. Often g_{ob} is negligible and the correction need not be used. The output capacity of the transistor will be lumped in parallel with L, thus making $y_{ob} = 0$ in the analysis. The reverse transfer admittance y_{rb} will be neglected to simplify the analysis.

The network which must be represented by its Z-parameters is given in Figure C-3.

From this circuit, the Z-parameters can be calculated using the following definitions:

Input impedance
$$Z_i = \frac{V_1}{I_1}\bigg|_{I_2 = 0} \tag{C-1}$$

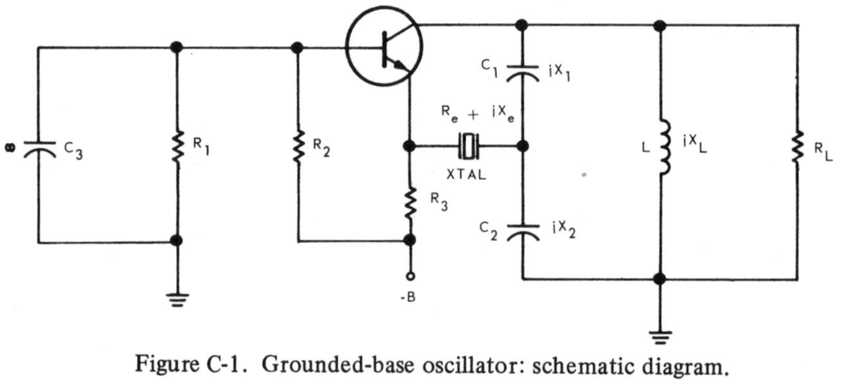

Figure C-1. Grounded-base oscillator: schematic diagram.

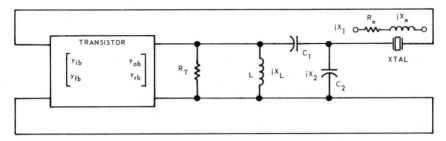

Figure C-2. Grounded-base oscillator: ac circuit diagram.

Output impedance
$$Z_o = \frac{V_2}{I_2}\bigg|_{I_1 = 0} \qquad (C\text{-}2)$$

Forward transfer impedance
$$Z_f = \frac{V_2}{I_1}\bigg|_{I_2 = 0} \qquad (C\text{-}3)$$

Reverse transfer impedance
$$Z_r = \frac{V_1}{I_2}\bigg|_{I_1 = 0} . \qquad (C\text{-}4)$$

To simplify the analysis, it will be assumed that $X_1 + X_2 + X_L = 0$. This is approximately resonance for the tank circuit.

From equation (C-1) it can be seen that Z_i is merely the input impedance of the network with the output open-circuited and can be written by inspection as $Z_i = R_T$.

The forward transfer impedance is given by the ratio of V_2 to I_1 with the output open-circuited. In order to calculate this, it is convenient to first deter-

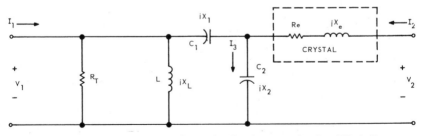

Figure C-3. Grounded-base oscillator feedback network: simplified diagram.

mine I_3 in terms of I_1, which is given by

$$I_3 = \frac{I_1\left(\dfrac{jR_T X_L}{R_T + jX_L}\right)}{\dfrac{jR_T X_L}{R_T + jX_L} + j(x_1 + x_2)}. \tag{C-5}$$

Simplifying gives

$$I_3 = \frac{jI_1 R_T X_L}{-X_L(X_1 + X_2) + jR_T(X_1 + X_2 + X_L)}, \tag{C-6}$$

but

$$I_3 = \frac{V_2}{jX_2}.$$

Substituting this and making use of the assumption that

$$X_L + X_1 + X_2 = 0$$

gives

$$\frac{V_2}{I_1} = \frac{R_T X_2 X_L}{X_L(X_1 + X_2)} = \frac{R_T X_2}{X_1 + X_2} = \frac{-R_T X_2}{X_L}; \tag{C-7}$$

therefore

$$Z_f = -\frac{R_T X_2}{X_L}.$$

Since the network is composed entirely of reciprocal elements, the forward and reverse transfer impedances are equal, so that

$$Z_f = Z_r = -\frac{R_T X_2}{X_L}. \tag{C-8}$$

From equation (C-2), it can be seen that Z_o is merely the output impedance of the network with the input open-circuited. It is given by

$$Z_o = R_e + jX_e + \frac{\left[\left(\dfrac{jR_T X_L}{R_T + jX_L}\right) + jX_1\right] jX_2}{\left[\left(\dfrac{jR_T X_L}{R_T + jX_L}\right) + jX_1\right] + jX_2}. \tag{C-9}$$

Simplifying this gives

$$Z_o = R_e + jX_e + \frac{jX_2[-X_1 X_L + jR_T(X_1 + X_L)]}{jR_T(X_1 + X_2 + X_L) - X_L(X_1 + X_2)}. \tag{C-10}$$

Making use of the assumption $X_1 + X_2 + X_L = 0$ and further simplifying gives

$$Z_o = R_e + jX_e + \frac{R_T X_2 (X_1 + X_L) + jX_1 X_2 X_L}{X_L(X_1 + X_2)} \tag{C-11}$$

and

$$Z_o = R_e + jX_e + \frac{R_T X_2^2 - jX_1 X_2 X_L}{X_L^2}. \tag{C-12}$$

These results can be substituted in the general equation for oscillation as developed in Appendix A:

$$y_{fb} Z_f + y_{ib} Z_o + y_{ob} Z_i + y_{rb} Z_r + \Delta y_b \Delta Z + 1 = 0. \tag{C-13}$$

Since y_{ob} is being accounted for the feedback network, and since y_{rb} is assumed to be zero, $\Delta y = y_{ib} y_{ob} - y_{fb} y_{ob} = 0$. This simplifies equation (C-13) to

$$y_{fb} Z_f + y_{ib} Z_o + 1 = 0. \tag{C-14}$$

Substituting for Z_f and Z_o gives

$$\frac{-y_{fb} R_T X_2}{X_L} + y_{ib}\left[R_e + jX_e + \frac{R_T X_2^2 - jX_1 X_2 X_L}{X_L^2}\right] + 1 = 0. \tag{C-15}$$

Substituting $y_{fb} = g_{fb} + jb_{fb}$ and $y_{ib} = g_{ib} + jb_{ib}$ gives

$$\frac{-(g_{fb} + jb_{fb}) R_T X_2}{X_L} + (g_{ib} + jb_{ib})\left[R_e + jX_e + \frac{R_T X_2^2 - jX_1 X_2 X_L}{X_L^2}\right] + 1 = 0.$$

$$\tag{C-16}$$

Performing the indicated multiplications and collecting terms results in the following equation:

$$-g_{fb} R_T X_2 - jb_{fb} R_T X_2 + g_{ib} R_e X_L + jg_{ib} X_e X_L + jb_{ib} R_e X_L - b_{ib} X_e X_L$$

$$+ \frac{g_{ib} R_T X_2^2}{X_L} - jg_{ib} X_1 X_2 + \frac{jR_T X_2^2 b_{ib}}{X_L} + b_{ib} X_1 X_2 + X_L = 0. \tag{C-17}$$

This equation can be separated into real and imaginary components. The real parts are

$$-g_{fb}R_TX_2 + g_{ib}R_eX_L - b_{ib}X_eX_L + \frac{g_{ib}R_TX_2^2}{X_L} + b_{ib}X_1X_2 + X_L = 0.$$

(C-18)

Substituting $g_{ib} = 1/R_{in}$ and simplifying gives

$$g_{fb} = \frac{1}{R_T}\left(\frac{X_L}{X_2}\right)\left[\frac{R_e + R_{in}}{R_{in}}\right] + \frac{1}{R_{in}}\left(\frac{X_2}{X_L}\right) + \frac{b_{ib}X_1}{R_T} - b_{ib}\left(\frac{X_e}{R_T}\right)\left(\frac{X_L}{X_2}\right) = 0.$$

(C-19)

The imaginary parts are

$$-b_{fb}R_TX_2 + g_{ib}X_eX_L + b_{ib}R_eX_L - g_{ib}X_1X_2 + \frac{R_TX_2^2 b_{ib}}{X_L} = 0. \quad \text{(C-20)}$$

Again substituting $g_{ib} = 1/R_{in}$ and simplifying gives

$$X_e = b_{fb}R_TR_{in}\left(\frac{X_2}{X_L}\right) + X_1\left(\frac{X_2}{X_L}\right) - b_{ib}R_{in}\left[R_e + R_T\left(\frac{X_2}{X_L}\right)^2\right]. \quad \text{(C-21)}$$

The optimum value of X_L/X_2 with respect to transistor gain can be found by differentiating equation (C-19):

$$\frac{d(g_{fb})}{d(X_L/X_2)} = \frac{1}{R_T}\left[\frac{R_e + R_{in}}{R_{in}}\right] - \frac{1}{R_{in}}\left(\frac{X_2}{X_L}\right)^2 - b_{ib}\left(\frac{X_e}{R_T}\right). \quad \text{(C-22)}$$

If it is required that the crystal operate at series resonance $X_e = 0$ or if b_{ib} is negligible, then equation (C-22) simplifies to

$$\frac{d(g_{fb})}{d(X_L/X_2)} = \frac{1}{R_T}\left[\frac{R_e + R_{in}}{R_{in}}\right] - \frac{1}{R_{in}}\left(\frac{X_2}{X_L}\right)^2. \quad \text{(C-23)}$$

Solving this for X_L/X_2 gives

$$X_L/X_2 = -\sqrt{R_T/(R_e + R_{in})} \quad \text{(C-24)}$$

The minimum g_{fb} then is given by

$$\left|g_{fb}\right|_{min} = \left|-\frac{2}{R_{in}}\sqrt{(R_e + R_{in})/R_T} + b_{ib}(X_1/R_T)\right|. \quad \text{(C-25)}$$

NOTE. This assumes that the crystal is at series resonance, $X_e = 0$.

Appendix D
Derivation of Approximate Equations for the Clapp Oscillator

The impedance Z_L, can be written as

$$Z_L = \frac{(R_e + jX_e)\left[jX_2 + \dfrac{(1/g_m)\,(jX_1)}{(1/g_m) + jX_1}\right]}{R_e + jX_e + jX_2 + \dfrac{(1/g_m)\,(jX_1)}{(1/g_m) + jX_1}}.$$

This can be simplified to

$$Z_L = \frac{(R_e + jX_e)\,[jX_2(1 + jX_1 g_m) + jX_1]}{(R_e + jX_e + jX_2)\,(1 + jX_1 g_m) + jX_1}.$$

It can be further rearranged to the form

$$Z_L = \frac{(R_e + jX_e)\,[j(X_1 + X_2) - X_1 X_2 g_m]}{R_e - X_e X_1 g_m - X_1 X_2 g_m + j(X_1 + X_2 + X_e) + jR_e X_1 g_m}.$$

If we now assume that

$$g_m X_1 X_2 \ll X_e, \qquad X_1 + X_2 + X_e = 0, \qquad \text{and} \qquad R_e \ll X_e,$$

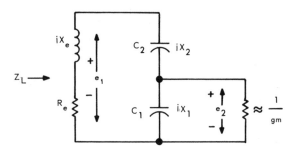

Figure D-1. Clapp oscillator tank circuit: simplified diagram.

then the expression may be simplified to

$$Z_L = \frac{-X_e(X_1 + X_2)}{R_e - g_m X_1(X_e + X_2)};$$

but since $X_1 + X_2 + X_e = 0$,

$$Z_L = \frac{(X_1 + X_2)^2}{R_e + g_m X_1^2}.$$

The voltage ratio e_2/e_1 may be found to be

$$\frac{e_2}{e_1} = \frac{\dfrac{(1/g_m)\,(jX_1)}{(1/g_m) + jX_1}}{jX_2 + \dfrac{(1/g_m)\,jX_1}{(1/g_m) + jX_1}}.$$

This simplifies to

$$\frac{e_2}{e_1} = \frac{X_1}{X_1 + X_2 + jX_1 X_2 g_m}.$$

If we assume that $X_1 + X_2 \gg X_1 X_2 g_m$, then the expression simplifies to

$$\frac{e_2}{e_1} \doteq \frac{X_1}{X_1 + X_2}.$$

Appendix E

Derivation of Approximate Equations for the Pierce Oscillator Analysis

The input impedance Z_L may be written as

$$Z_L = \frac{jX_2(R_e + jX_e + jX_1)}{R_e + jX_2 + jX_1 + jX_e}.$$

If $X_1 + X_2 + X_e = 0$, then the expression simplifies to

$$Z_L = \frac{-X_2(X_e + X_1) + jR_eX_2}{R_e}.$$

Again applying the preceding assumption,

$$Z_L = \frac{X_2}{R_e}(X_2 + jR_e).$$

If we now assume that $X_2 \gg R_e$, the input impedance simplifies to

$$Z_L = \frac{X_2^2}{R_e}.$$

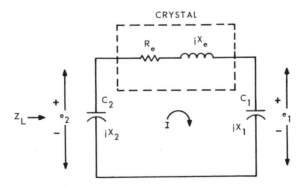

Figure E-1. Pierce oscillator π-network: simplified diagram.

The voltage e_1 may be written as $e_1 = jX_1I$, where

$$I = \frac{e_2}{R_e + j(X_e + X_1)}.$$

Combining these gives

$$\frac{e_1}{e_2} = \frac{jX_1}{R_e + j(X_1 + X_e)}.$$

If $X_1 + X_2 + X_e = 0$, then this expression simplifies to

$$\frac{e_1}{e_2} = \frac{jX_1}{R_e - jX_2}.$$

If now we assume that $X_2 \gg R_e$, then

$$\frac{e_1}{e_2} = -\frac{X_1}{X_2}.$$

Also, since

$$X_1 = -\frac{1}{\omega C_1}$$

and

$$X_2 = -\frac{1}{\omega C_2},$$

then

$$\frac{e_1}{e_2} = -\frac{C_2}{C_1}.$$

Appendix F

Derivation of Approximate Equations for the Colpitts Oscillator

The input impedance can be written as

$$Z_L = \frac{jX_2(R_e + jX_1 + jX_e)}{jX_2 + jX_1 + jX_e + R_e}.$$

Assuming that $X_1 + X_2 + X_e = 0$ gives

$$Z_L = \frac{-X_2(X_1 + X_e) + jR_eX_2}{R_e}.$$

Again applying the preceding assumption gives

$$Z_L = \frac{X_2(X_2 + jR_e)}{R_e}.$$

If we now assume that $X_2 \gg R_e$, we have

$$Z_L = \frac{X_2^2}{R_e}.$$

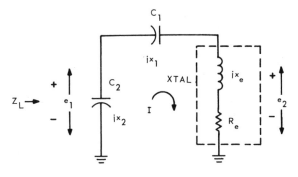

Figure F-1. Colpitts oscillator phase shift circuit: simplified diagram.

The voltage e_2 may be written as

$$e_2 = I(R_e + jX_e),$$

where

$$I = \frac{e_1}{jX_1 + R_e + jX_e}.$$

Combining these gives

$$\frac{e_2}{e_1} = \frac{R_e + jX_e}{R_e + jX_1 + jX_e}.$$

If we again assume that $X_1 + X_2 + X_e = 0$, then $X_e = -(X_1 + X_2)$ and $X_1 = -(X_2 + X_e)$. Substituting these gives

$$\frac{e_2}{e_1} = \frac{R_e - j(X_1 + X_2)}{R_e - jX_2}.$$

Assuming now that $R_e \ll X_2$ and $R_e \ll X_1$, we have

$$\frac{e_2}{e_1} = \frac{X_1 + X_2}{X_2}.$$

Appendix G
Large-Signal Transistor Parameters

In general, a transistor can be thought of as being made up of intrinsic and ex-trinsic elements. The extrinsic elements in general result from resistance, capaci-tance, or inductance in the leads connected to the semiconductor material. The bulk resistances of the semiconductor material also give rise to elements which to a first approximation may be lumped into the extrinsic elements. Thus, the transistor may be represented as shown in Figure G-1.

In many applications, particularly when the emitter current is low and the fre-quency considerably below f_t, the circuit behavior is not appreciably affected by the extrinsic elements. It is then possible to consider only the behavior of the intrinsic transistor, which is done for this analysis.

Referring to Figure G-1, it is known that the intrinsic base voltage is given by

$$V_{b'} = \frac{\lambda KT}{q} \ln \left[1 + \frac{I_e - \alpha_I I_c}{I_{ed}} \right]. \qquad \text{(G-1)}$$

Most of the symbols have been previously defined in this book; therefore, a complete explanation is not given here. Since λ is usually near unity and to sim-

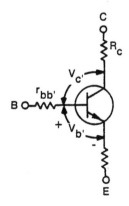

Figure G-1. Transistor model showing extrinsic elements.

plify the terminology, λ is taken to be unity for the analysis. It may be reinserted later if desired by replacing K by λK in the equations.

The term α_I is the inverted current gain I_e/I_c with the collector acting as emitter. I_{ed} is the diffusion saturation current. It should be pointed out that I_{ed} is very much a function of temperature and, in general, doubles about every $10°C$. This effect results in a larger change in $V_{b'}$ with temperature than the q/KT term. In the active region the collector current is given by

$$I_c = \alpha_N I_e \tag{G-2}$$

where α_N is the normal forward emitter-to-collector current gain. This equation is used in equation (G-1) in both the active region and for large signals in the cutoff region during part of each cycle. While equation (G-2) does not necessarily hold during cutoff, the collector current is so small during that part of the cycle that a considerable error has very little effect on the overall behavior. Making the indicated substitution yields the equation

$$V_{b'} = \frac{KT}{q} \ln \left[1 + \frac{I_e(1 - \alpha_I \alpha_N)}{I_{ed}} \right]. \tag{G-3}$$

Solving for I_e gives

$$I_e = \frac{I_{ed}}{(1 - \alpha_I \alpha_N)} [e^{qV_{b'}/KT} - 1]. \tag{G-4}$$

Now let

$$I_R = \frac{I_{ed}}{(1 - \alpha_I \alpha_N)}. \tag{G-5}$$

Substituting this into equation (G-4) gives

$$I_e = I_R (e^{qV_{b'}/KT} - 1), \tag{G-6}$$

which is in the form of the well known Shockley equation for an ideal p-n junction. I_R is generally in the order of three times the value of I_{ed} and thus is a very small current. Therefore, for practical purposes, equation (G-6) is equivalent to equation (G-7) if the transistor is active during at least a part of each cycle:

$$I_e = I_R e^{qV_{b'}/KT} \tag{G-7}$$

We now assume that a sinusoidal signal is applied between the base and emitter. In general, some dc bias is also applied and the base-to-emitter voltage has the form

$$V_{b'} = E \cos \omega t + E_0. \tag{G-8}$$

The emitter current is then given by

$$I_e = I_R \, \exp\left[\frac{q}{KT}(E \cos \omega t + E_0)\right]. \tag{G-9}$$

This can be rewritten in the form:

$$I_e = I_R \, \exp\left(\frac{qE_0}{KT}\right) \exp\left(\frac{q}{KT}E \cos \omega t\right). \tag{G-10}$$

Let us now examine the term

$$\exp\left(\frac{q}{KT}E \cos \omega t\right). \tag{G-11}$$

To expand this, we use the Bessel function expansion of the form*

$$\exp z \cos \theta = I_0(z) + 2 \sum_{n=1}^{\infty} I_n(z) \cos n\theta, \tag{G-12}$$

where $I_n(z)$ represents a modified Bessel function of the first kind and of order n. These modified Bessel functions are also referred to as the hyperbolic Bessel functions and are related to the familiar $J_n(z)$ Bessel functions much as the trigonometric functions are related to the hyperbolic functions. Thus,

$$I_n(z) = (i^{-n}) J_n(iz), \tag{G-13}$$

where

$$i = \sqrt{-1}.$$

In series form,

$$I_n(z) = \sum_{j=0}^{\infty} \frac{(z/2)^{n+2j}}{j!(n+j)!}. \tag{G-14}$$

Substituting $z = qE/KT$ into equation (G-12) and substituting equation (G-12) into equation (G-10), we have

$$I_e = I_R \, \exp\left(\frac{qE_0}{KT}\right)\left[I_0\left(\frac{Eq}{KT}\right) + 2 \sum_{n=1}^{\infty} I_n\left(\frac{Eq}{KT}\right) \cos n\omega t\right]. \tag{G-15}$$

For convenience in notation, let $V = Eq/KT$. Substituting this into equaion (G-15) and writing out a few terms, we have

$$I_e = I_R(\exp qE_0/KT)[I_0(V) + 2I_1(V) \cos \omega t + 2I_2(V) \cos 2\omega t + \cdots]. \tag{G-16}$$

*See page 106 of reference 33.

The dc component is given by

$$I_e(\text{mean}) = I_R \exp qE_0/KT\, I_0(V). \qquad \text{(G-17)}$$

Substituting this into equation (G-16) to eliminate E_0, we have

$$I_e = I_e(\text{mean})\left[1 + \frac{2I_1(V)}{I_0(V)}\cos \omega t + \frac{2I_2(V)}{I_0(V)}\cos 2\omega t + \cdots\right]. \qquad \text{(G-18)}$$

As stated earlier, $I_c = \alpha_N I_e$, but α_N is nearly unity; hence, $I_c \doteq I_e$. To determine the transconductance, we form the ratio,

$$g_m = \frac{i_c(\text{fund})}{E\cos \omega t}, \qquad \text{(G-19)}$$

which is given by

$$g_m = \frac{2I_e(\text{mean})\, I_1(V)\cos \omega t}{I_0(V)\, E\cos \omega t} = \frac{2I_e(\text{mean})\, I_1(V)}{E I_0(V)}. \qquad \text{(G-20)}$$

In the small-signal case, $V = 0$ and $I_0(V) = 1$, while

$$I_1(V) = \frac{V}{2} = \frac{Eq}{2KT}. \qquad \text{(G-21)}$$

Substituting these in equation (G-20) gives

$$g_{mo} = \frac{2I_e(\text{mean})\, Eq}{2KTE} = \frac{I_e(\text{mean})\, q}{KT}, \qquad \text{(G-22)}$$

which is the well known small-signal value. Substituting this in equation (G-20) for $I_e(\text{mean})$ gives

$$\frac{g_m}{g_{mo}} = \frac{2I_1(V)}{VI_0(V)}. \qquad \text{(G-23)}$$

It is also possible to determine the input impedance of the transistor from equation (G-18) making use of the fact that the base current is related to the emitter current by the factor $\beta + 1$; hence,

$$I_b = \frac{I_e}{\beta + 1}. \qquad \text{(G-24)}$$

The equivalent input resistance is given by

$$R_{\text{in}} = \frac{E\cos \omega t}{I_b(\text{fund})}, \qquad \text{(G-25)}$$

which is

$$R_{in} = \frac{E \cos \omega t \, (\beta + 1) I_0(V)}{2 I_e(\text{mean}) I_1(V) \cos \omega t} = \frac{E(\beta + 1) I_0(V)}{2 I_e(\text{mean}) I_1(V)}. \tag{G-26}$$

In the small-signal case $V = 0$ and, using the limit values of equation (G-21), we have

$$R_{ino} = \frac{E(\beta + 1)}{2 I_e(\text{mean}) \, (Eq/2KT)} = \frac{(\beta + 1) KT}{q I_e}, \tag{G-27}$$

which is the well known small-signal value. Substituting this into equation (G-26) gives

$$\frac{R_{in}}{R_{ino}} = \frac{V I_0(V)}{2 I_1(V)}. \tag{G-28A}$$

For purposes of computer analysis it is possible to approximate equation (G-28A) by the simplified equation

$$\frac{r_e}{r_{eo}} = \left[1 + \left(\frac{Eq}{2KT\lambda} \right)^2 \right]^{1/2} = \left[1 + \left(\frac{5.78 \times 10^{-3} E}{T\lambda} \right)^2 \right]^{1/2}. \tag{G-28B}$$

The diffusion capacitance for large-signal voltages may also be computed. The charge stored in the base region is proportional to the base current; hence

$$Q = M I_b, \tag{G-29}$$

where M is a constant of proportionality related to the carrier lifetime. Substituting equation (G-24) into (G-7) and the resultant equation into (G-29) gives

$$Q = \frac{M I_R}{(\beta + 1)} e^{q V_{b'}/KT}. \tag{G-30}$$

The reactive component of the base current is given by

$$i = \frac{dQ}{dt}. \tag{G-31}$$

Differentiating equation (G-30) and substituting in to (G-31) gives

$$i = \frac{M I_R}{(\beta + 1)} \frac{q}{KT} e^{q V_{b'}/KT} \frac{dV_{b'}}{dt}. \tag{G-32}$$

Substituting $V_{b'} = E \cos \omega t + E_0$ gives

$$i = \frac{-M I_R q E \omega}{(\beta + 1) KT} \exp\left(\frac{q E_0}{KT} \right) \left[\exp\left(\frac{q E}{KT} \cos \omega t \right) \right] \sin \omega t. \tag{G-33}$$

For convenience in notation, let

$$P = -\frac{MI_R qE\omega}{(\beta + 1) KT} \exp\left(\frac{qE_0}{KT}\right); \tag{G-34}$$

then

$$i = P\left(\exp\frac{qE \cos \omega t}{KT}\right) \sin \omega t. \tag{G-35}$$

The fundamental component of this current may be found by Fourier analysis using the formulas given in equations (H-5), (H-6), and (H-7) of Appendix H. Thus the coefficient of the sin ωt term is given by

$$b_1 = \frac{1}{\pi} \int_0^{2\pi} i \sin \theta \, d\theta, \tag{G-36}$$

where $\theta = \omega t$. Substituting for i from equation (G-35) gives

$$b_1 = \frac{1}{\pi} \int_0^{2\pi} P\left(\exp\frac{qE \cos \theta}{KT}\right) \sin \theta \sin \theta \, d\theta. \tag{G-37}$$

Using the identity, $\sin^2 \theta = \frac{1}{2} - \frac{1}{2} \cos 2\theta$,

$$b_1 = \frac{1}{2\pi} \int_0^{2\pi} P\left(\exp\frac{qE \cos \theta}{KT}\right) d\theta$$

$$-\frac{1}{2\pi} \int_0^{2\pi} P\left(\exp\frac{qE \cos \theta}{KT}\right) \cos 2\theta \, d\theta. \tag{G-38}$$

The first integral may be evaluated using the form*

$$\frac{1}{2\pi} \int_0^{2\pi} (\exp z \cos \theta) \, d\theta = I_0(Z), \tag{G-39}$$

and the second, using the form**

$$\frac{1}{2\pi} \int_0^{2\pi} (\exp z \cos \theta) \cos n\theta \, d\theta = (-1)^n I_n(z).$$

*See page 162 of reference 33.
**See page 51 of reference 33.

Using these forms, we have

$$b_1 = P\left[I_0\left(\frac{qE}{KT}\right) - I_2\left(\frac{qE}{KT}\right)\right].$$ (G-40)

It can be shown that[***]

$$I_0(z) - I_2(z) = \frac{2I_1(z)}{z};$$ (G-41)

therefore,

$$b_1 = \frac{2PI_1(qE/KT)}{qE/KT} = \frac{2PI_1(V)}{V}.$$ (G-42)

The coefficient of the cos ωt term is given by

$$a_1 = \frac{1}{\pi}\int_0^{2\pi} P\left(\exp\frac{qE\cos\theta}{KT}\right)\sin\theta\cos\theta\,d\theta.$$ (G-43)

Using the identity

$$\sin\theta\cos\theta = \tfrac{1}{2}\sin 2\theta,$$ (G-44)

we have

$$a_1 = \frac{1}{2\pi}\int_0^{2\pi} P\left(\exp\frac{qE\cos\theta}{KT}\right)\sin 2\theta\,d\theta.$$ (G-45)

The value of this integral can be shown to be zero.[*] Thus the fundamental component of capacitive current is given by

$$i = \frac{2PI_1(V)}{V}\sin\omega t.$$ (G-46)

The equivalent capacitance is given by the relationship,

$$\omega C = \frac{i(\text{fund})\,(90^\circ\text{ leading})}{e(\text{fund})}.$$ (G-47)

Now $-\sin\omega t$ leads cos ωt by 90 degrees; hence:

$$\omega C_{de} = \frac{-2PI_1(V)}{EV},$$ (G-48)

[***]See page 163 of reference 33.
 [*]See page 51 of reference 33.

since $e = E \cos \omega t$. Substituting for P from equation (G-34),

$$C_{de} = \frac{Mq}{KT(\beta + 1)} \frac{2I_1(V)}{V} I_R\, e^{qE_0/KT}.$$ (G-49)

Substituting equation (G-17) for $I_R \exp qE_0/KT$ gives

$$C_{de} = \frac{M}{\beta + 1} \frac{qI_e(\text{mean})}{KT} \frac{2I_1(V)}{VI_0(V)}.$$ (G-50)

For the small-signal case, we may use the limit values given in (G-21) and

$$C_{deo} = \frac{M}{\beta + 1} \frac{qI_e(\text{mean})}{KT}.$$ (G-51)

Therefore, from the circuit of Figure 6-5 and equation (6-16),

$$\frac{M}{\beta + 1} = \frac{1}{2\pi f_t}.$$ (G-52)

Also dividing equation (G-50) by equation (G-51), we have

$$\frac{C_{de}}{C_{deo}} = \frac{2I_1(V)}{VI_0(V)}.$$ (G-53)

The bias shift resulting from the signal voltage can be determined from equation (G-17). With signal present,

$$I_e(\text{mean}) = I_R(\exp qE_0/KT)\, I_0(V).$$ (G-54)

With no signal, $V = 0$ and $I_0(V) = 1$; hence,

$$I_e(\text{no signal}) = I_R \exp qV_{be}/KT.$$ (G-55)

We wish to determine E_0 so that

$$I_e(\text{mean}) = I_e(\text{no signal}).$$

Equating gives

$$(\exp qE_0/KT)\, I_0(V) = \exp qV_{be}/KT.$$ (G-56)

Solving for qE_0/KT gives

$$\frac{qE_0}{KT} = \frac{qV_{be}}{KT} - \ln I_0(V),$$ (G-57)

or

$$E_0 - V_{be} = \frac{KT}{q} \ln I_0(V).$$ (G-58)

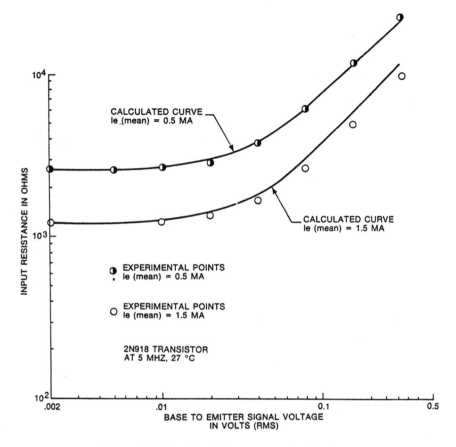

Figure G-2. Input resistance versus signal voltage.

The bias shift is given by

$$V_{bias} = E_0 - V_{be} = \frac{KT}{q} \ln I_0(V) \qquad (G\text{-}59)$$

Noting that $V = qE/KT$, we have

$$V_{bias} = \frac{KT}{q} \ln I_0\left(\frac{qE}{KT}\right). \qquad (G\text{-}60)$$

In order to verify the results of this analysis, a series of measurements was made on a type 2N918 transistor. Tests were run at emitter currents of both 0.5 mA and at 1.5 mA. The ac sinusoidal input voltage was varied from 0.002 to

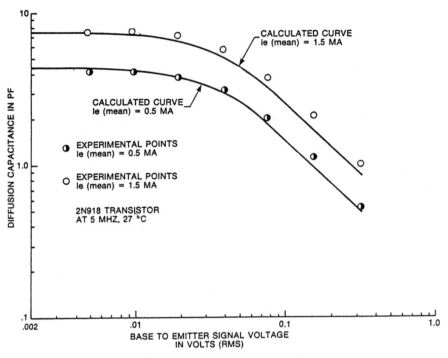

Figure G-3. Diffusion capacitance versus signal voltage.

0.3 V. As the signal voltage was increased, the bias circuit was readjusted to maintain the emitter current at the indicated value. The resistance and capacitance measurements were made, using a Boonton RX meter, by adjusting the oscillator to obtain the desired ac voltage. The graph of Figure G-2 shows the input resistance variation as a function of signal level, while Figure G-3 shows how the input capacitance varies with drive. The curves are plotted from equation (G-28B), and the circles correspond to measured points. As can be seen, the agreement is excellent.

Appendix H
Large-Signal Transistor Parameters with Emitter Degeneration

The emitter current of a transistor in the common-emitter configuration has a characteristic similar to that shown in Figure H-1. If a significant amount of emitter degeneration is used, the curve may be approximated by a straight line as shown. The slope of this line is $1/(r_e + R_f)_0$ where r_e is the small-signal transistor emitter resistance calculated at the mean emitter current I_e, and R_f is the external emitter degeneration resistance. Thus the emitter current is given approximately by

$$i_e = \frac{E \sin \omega t + E_0}{(r_e + R_f)_0} \qquad (E \sin \omega t + E_0) \geqslant 0 \qquad \text{(H-1)}$$

$$i_e = 0 \qquad (E \sin \omega t + E_0) < 0. \qquad \text{(H-2)}$$

The collector current is given by $i_c = \alpha i_e$. Normally α is very near unity and the collector is taken to be equal to the emitter current for this analysis.

The effective conduction angle is 2θ, and the voltage E_0 is given by

$$E_0 = -E \cos \theta. \qquad \text{(H-3)}$$

Thus we may write

$$i_e = \frac{E \sin \omega t - E \cos \theta}{(r_e + R_f)_0} \qquad \left(\frac{\pi}{2} - \theta\right) \leqslant \omega t \leqslant \left(\frac{\pi}{2} + \theta\right) \qquad \text{(H-4)}$$

$$i_e = 0 \qquad \text{elsewhere.}$$

Using Fourier analysis, we may represent i_e by an infinite series of the form

$$i = \frac{a_0}{2} + \sum_{n=1}^{\infty} a_n \cos n\omega t + b_n \sin n\omega t, \qquad \text{(H-5)}$$

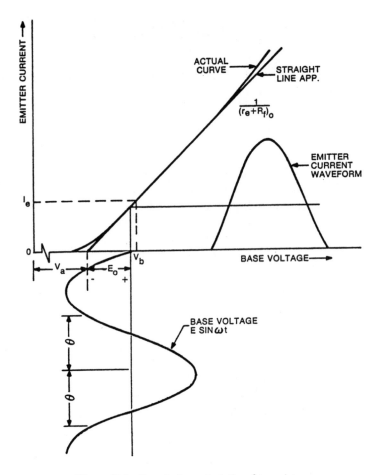

Figure H-1. Input characteristic of transistor.

where the a_n and b_n terms are defined by the equations

$$a_n = \frac{1}{\pi} \int_0^{2\pi} i_e \cos n\phi \, d\phi \qquad \text{(H-6)}$$

$$b_n = \frac{1}{\pi} \int_0^{2\pi} i_e \sin n\phi \, d\phi, \qquad \text{(H-7)}$$

where $\phi = \omega t$. Let us first find the dc value of the current given by $a_0/2$.

$$a_0 = \frac{1}{\pi} \int_{(\pi/2)-\theta}^{(\pi/2)+\theta} \frac{E \sin \phi - E \cos \theta}{(r_e + R_f)_0} \, d\phi$$

$$= \frac{2E (\sin \theta - \theta \cos \theta)}{(R_f + r_e)\pi}. \tag{H-8}$$

Thus the dc emitter current is given by

$$I_e(\text{mean}) = \frac{E}{(r_e + R_f)_0 \pi} (\sin \theta - \theta \cos \theta). \tag{H-9}$$

The fundamental component of current is given by the a_1 and b_1 terms, which are

$$a_1 = \frac{1}{\pi} \int_{(\pi/2)-\theta}^{(\pi/2)+\theta} \frac{E \sin \phi - E \cos \theta}{(r_e + R_f)_0} \cos \phi \, d\phi = 0 \tag{H-10}$$

$$b_1 = \frac{1}{\pi} \int_{(\pi/2)-\theta}^{(\pi/2)+\theta} \frac{E \sin \phi - E \cos \theta}{(r_e + R_f)_0} \sin \phi \, d\phi$$

$$= \frac{E(\theta - \frac{1}{2} \sin 2\theta)}{\pi(r_e + R_f)_0}. \tag{H-11}$$

The second harmonic component is given by the a_2 and b_2 terms, which are:

$$a_2 = \frac{1}{\pi} \int_{(\pi/2)-\theta}^{(\pi/2)+\theta} \frac{E \sin \phi - E \cos \theta}{(r_e + R_f)_0} \cos 2\phi \, d\phi$$

$$= \frac{E(-\frac{1}{3} \sin 3\theta - \sin \theta + \cos \theta \sin 2\theta)}{\pi(r_e + R_f)_0} \tag{H-12}$$

$$b_2 = \frac{1}{\pi} \int_{(\pi/2)-\theta}^{(\pi/2)+\theta} \frac{E \sin \phi - E \cos \theta}{(r_e + R_f)_0} \sin 2\phi \, d\phi = 0. \tag{H-13}$$

The third harmonic component is given by the a_3 and b_3 terms, which are

$$a_3 = \frac{1}{\pi} \int_{(\pi/2)-\theta}^{(\pi/2)+\theta} \frac{E \sin \phi - E \cos \theta}{(r_e + R_f)_0} \cos 3\phi \, d\phi = 0. \tag{H-14}$$

$$b_3 = \frac{1}{\pi} \int_{(\pi/2)-\theta}^{(\pi/2)+\theta} \frac{E \sin \phi - E \cos \theta}{(r_e + R_f)_0} \sin 3\phi \, d\phi$$

$$= \frac{E(-\frac{1}{2} \sin 2\theta + \frac{2}{3} \cos \theta \sin 3\theta - \frac{1}{4} \sin 4\theta)}{\pi(r_e + R_f)_0}. \tag{H-15}$$

Substituting these terms in equation (H-5) gives the equation

$$i = \frac{E(\sin \theta - \theta \cos \theta)}{\pi(r_e + R_f)_0} + \frac{E(\theta - \frac{1}{2} \sin 2\theta)}{\pi(r_e + R_f)_0} \sin \omega t$$

$$+ \frac{E(-\frac{1}{3} \sin 3\theta - \sin \theta + \cos \theta \sin 2\theta)}{\pi(r_e + R_f)_0} \cos 2\omega t$$

$$+ \frac{E(-\frac{1}{2} \sin 2\theta + \frac{2}{3} \cos \theta \sin 3\theta - \frac{1}{4} \sin 4\theta)}{\pi(r_e + R_f)_0} \sin 3\omega t + \cdots. \tag{H-16}$$

Substituting for $(r_e + R_f)_0$ from equation (H-9),

$$i = I_e(\text{mean}) \left[1 + \frac{(\theta - \frac{1}{2} \sin 2\theta)}{(\sin \theta - \theta \cos \theta)} \sin \omega t \right.$$

$$+ \frac{(-\frac{1}{3} \sin 3\theta - \sin \theta + \cos \theta \sin 2\theta)}{(\sin \theta - \theta \cos \theta)} \cos 2\omega t$$

$$\left. + \frac{(-\frac{1}{2} \sin 2\theta + \frac{2}{3} \cos \theta \sin 3\theta - \frac{1}{4} \sin 4\theta)}{(\sin \theta - \theta \cos \theta)} \sin 3\omega t + \cdots \right]. \tag{H-17}$$

Thus it can be seen that knowing the mean emitter current and the conduction angle, the fundamental and harmonic components of the collector current may be calculated. The conduction angle is found using equation (H-9). Unfortunately, θ cannot be solved for directly, but the equations can be plotted in parametric form using θ as the parameter. This was done and the results are presented in section 6.4.2.

The equivalent input impedance of the transistor is given by

$$R_{\text{in}} = \frac{(\text{fundamental component of base voltage})}{(\text{fundamental component of base current})}. \tag{H-18}$$

The base current is given by $i_e/(\beta + 1)$; therefore, from equation (H-17) we have

$$i_b(\text{fund}) = \frac{I_e(\text{mean}) (\theta - \frac{1}{2} \sin 2\theta)}{(\beta + 1) (\sin \theta - \theta \cos \theta)} \sin \omega t. \tag{H-19}$$

Substituting $E \sin \omega t$ for the base voltage and equation (H-19) for the base current, the input resistance may be computed from equation (H-18) as

$$R_{in} = \frac{E(\beta + 1)(\sin \theta - \theta \cos \theta) \sin \omega t}{I_e(\text{mean})(\theta - \frac{1}{2}\sin 2\theta) \sin \omega t}$$

$$R_{in} = \frac{E(\beta + 1)(\sin \theta - \theta \cos \theta)}{I_e(\text{mean})(\theta - \frac{1}{2}\sin 2\theta)}. \tag{H-20}$$

Substituting for $I_e(\text{mean})$ from equation (H-9) gives

$$R_{in} = \frac{(\beta + 1)(r_e + R_f)_0 \pi}{(\theta - \frac{1}{2}\sin 2\theta)}.$$

If we define

$$\frac{r_e + R_f}{(r_e + R_f)_0} = \frac{\pi}{\theta - \frac{1}{2}\sin 2\theta}, \tag{H-21}$$

then

$$R_{in} = (\beta + 1)(r_e + R_f), \tag{H-22}$$

but the small-signal input resistance $(\theta = \pi)$ is given by

$$R_{ino} = (\beta + 1)(r_e + R_f)_0;$$

hence

$$\frac{R_{in}}{R_{ino}} = \frac{\pi}{\theta - \frac{1}{2}\sin 2\theta}. \tag{H-23}$$

The effective transconductance of the transistor is given by

$$g_m = \frac{\text{fundamental component of collector current}}{\text{fundamental component of base voltage}}. \tag{H-24}$$

Substituting the fundamental component of collector current from equation (B-17) and $E \sin \omega t$ for the base voltage, we have

$$g_m = \frac{I_e(\text{mean})(\theta - \frac{1}{2}\sin 2\theta) \sin \omega t}{E(\sin \theta - \theta \cos \theta) \sin \omega t}. \tag{H-25}$$

Substituting for the mean emitter current from equation (H-9), we have

$$g_m = \frac{(\theta - \frac{1}{2}\sin 2\theta)}{(r_e + R_f)_0 \pi}. \tag{H-26}$$

Again using the definition of equation (H-21), we have

$$g_m = \frac{1}{(r_e + R_f)},$$
(H-27)

but the small-signal transconductance is given by

$$g_{m0} = 1/(r_e + R_f)_0;$$

therefore,

$$\frac{g_m}{g_{m0}} = \frac{\theta - \frac{1}{2} \sin 2\theta}{\pi}.$$
(H-28)

It is also possible to calculate the bias shift due to the presence of the signal voltage. From Figure (H-1) it can be seen that a no-signal base voltage V_b is required to establish an emitter current I_e.

With a signal present and a desired I_e(mean) equal to the no-signal I_e, a voltage $V_a + E_0$, is required. By inspection we see that I_e(mean) $\times$ $(r_e + R_f)_0$ is the distance between the intersection of the resistance line with the X-axis and V_b. Thus $V_a = V_b - I_e$(mean) $(r_e + R_f)_0$. The bias required with signal is thus given by

$$V_B = V_a + E_0 = V_b - I_e(\text{mean}) (r_e + R_f)_0 + E_0.$$
(H-29)

But from the equation (H-3) we have

$$E_0 = -E \cos \theta;$$

therefore,

$$V_B = V_b - I_e(\text{mean}) (r_e + R_f)_0 - E \cos \theta.$$
(H-30)

Substituting for E from equation (H-9), we have

$$V_B = V_b - I_e(\text{mean}) (r_e + R_f)_0 \left(1 + \frac{\pi}{\tan \theta - \theta}\right).$$
(H-31)

The bias shift $V_B - V_b$ is therefore given by

$$V_{\text{bias}} = I_e(\text{mean}) (r_e + R_f)_0 \left[1 + \frac{\pi}{\tan \theta - \theta}\right].$$
(H-32)

Normalizing the shift to I_e(mean) $(r_e + R_f)_0$, we have

$$\frac{V_{\text{bias}}}{I_e(\text{mean}) (r_e + R_f)} = 1 + \frac{\pi}{\tan \theta - \theta}.$$
(H-33)

The peak current for $0 \leqslant \theta < \pi$ is given by

$$i(\text{peak}) = \frac{E + E_0}{(r_e + R_f)_0} = \frac{E(1 - \cos \theta)}{(r_e + R_f)_0}. \tag{H-34}$$

Thus using equation (H-9),

$$\frac{i(\text{peak})}{I_e(\text{mean})} = \frac{\pi(1 - \cos \theta)}{\sin \theta - \theta \cos \theta}. \tag{H-35}$$

For $\theta = \pi$,

$$i(\text{peak}) = I_e + \frac{E}{(r_e + R_f)_0}. \tag{H-36}$$

Appendix I

Nonlinear Analysis of the Colpitts Oscillator Based on the Principle of Harmonic Balance

A simplified schematic diagram of the Colpitts oscillator is shown below in Figure I-1. Here the crystal is represented by its series resistance R and an equivalent inductance L. For the analysis, it is assumed that R_1, R_2, and R_3 are large enough so that they produce only negligible effects on the RF signals. The transistor input and output capacitances are also neglected in this analysis. The effects of these reactances may be determined using the method given in sections 6.4.1 and 7.2.2.

From Appendix G, equation (G-7), we see that the emitter current of the transistor is given by

$$I_e = I_R \exp qV_b/KT \tag{I-1}$$

where I_R is a constant defined in equation (G-5) and V_b is the intrinsic base-to-emitter voltage. The base current is then given by $(\beta + 1)I_b = I_e$, where β is the common-emitter current gain. If we let $I_r = I_R/(\beta + 1)$, we may write

$$I_b = I_r \exp qV_b/KT \tag{I-2}$$

in the active region.*

The circuit of Figure I-1 is redrawn in Figure I-2 to show the base-to-emitter diode and the collector current generator. We note that $V_b = V_1 + V_0$, where V_0 is the base bias voltage.

The equations for the system can be written as follows:

$$V_2 - V_1 = R i_3 + L \frac{di_3}{dt} \tag{I-3}$$

$$i_2 = C_2 \frac{dV_2}{dt} \tag{I-4}$$

---

*Note that the "active region" here means any operating condition in which the transistor is not saturated ($V_{cb} > 0$) and is much greater than the linear region.

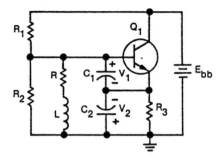

Figure I-1. Colpitts oscillator.

$$i_c + i_2 + i_3 = 0 \tag{I-5}$$

$$i_1 = C_1 \frac{dV_1}{dt} \tag{I-6}$$

$$i_3 = i_1 + I_b \tag{I-7}$$

$$I_b = I_r \exp qV_b/KT \tag{I-8}$$

$$I_c = \beta I_b \tag{I-9}$$

$$V_b = V_0 + V_1. \tag{I-10}$$

With considerable manipulation, these simultaneous equations can be used to solve for V_1. The resulting expression is

$$\frac{d^3 V_1}{dt^3} + \frac{d^2 V_1}{dt^2} \left\{ \frac{R}{L} + \frac{1}{C_1} \left(\frac{qI_r}{KT} \right) \exp \left[\frac{q}{KT} (V_1 + V_0) \right] \right\}$$
$$+ \left(\frac{dV_1}{dt} \right)^2 \left\{ \frac{I_r}{C_1} \left(\frac{q}{KT} \right)^2 \exp \left[\frac{q(V_1 + V_0)}{KT} \right] \right\}$$

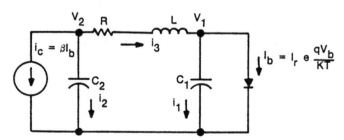

Figure I-2. Equivalent circuit of Colpitts oscillator (bias circuit not shown).

$$+ \frac{dV_1}{dt} \left\{ \frac{1}{L} \left(\frac{1}{C_1} + \frac{1}{C_2} \right) + \frac{R}{LC_1} \left(\frac{qI_r}{KT} \right) \exp \left[\frac{q}{KT} (V_1 + V_0) \right] \right\}$$

$$+ \frac{I_r(\beta + 1)}{LC_1 C_2} \exp \left[\frac{q}{KT} (V_1 + V_0) \right] = 0. \tag{I-11}$$

This equation is nonlinear not only because of the exponential terms but also because of the presence of the $(dV_1/dt)^2$ term. If this equation could be solved, it would exactly describe the behavior of the equivalent circuit from which it was derived. Unfortunately, an exact solution cannot be found. There are several techniques by which an approximate solution for this equation can be obtained. Perhaps the least difficult is the principle of harmonic balance. Using this technique, the voltage V_1 is assumed to have a solution of the form

$$V_1 = E \cos \omega t. \tag{I-12}$$

This expression is substituted for V_1 in the equation, and coefficients E and ω are adjusted so that the equation is exact insofar as terms of the fundamental frequency are concerned. Terms containing $\cos 2\omega t$, $\sin 2\omega t$, $\cos 3\omega t$, etc., are simply ignored. The justification for this rests in the theory that it is primarily the fundamental terms that determine the amplitude of oscillation and the frequency.

We see that if $V_1 = E \cos \omega t$,

$$\frac{dV_1}{dt} = -\omega E \sin \omega t \tag{I-13}$$

$$\frac{d^2 V_1}{dt^2} = -\omega^2 E \cos \omega t \tag{I-14}$$

$$\frac{d^3 V_1}{dt^3} = \omega^3 E \sin \omega t. \tag{I-15}$$

Making these substitutions, we obtain

$$\omega^3 E \sin \omega t - \omega^2 E \cos \omega t \left[\frac{R}{L} + \frac{1}{C_1} \left(\frac{qI_r}{KT} \right) \exp \left(\frac{qV_0}{KT} \right) \exp \left(\frac{qE \cos \omega t}{KT} \right) \right]$$

$$+ \omega^2 E^2 \sin^2 \omega t \left[\frac{I_r}{C_1} \left(\frac{q}{KT} \right)^2 \exp \left(\frac{qV_0}{KT} \right) \exp \left(\frac{qE \cos \omega t}{KT} \right) \right]$$

$$- \omega E \sin \omega t \left[\frac{1}{L} \left(\frac{1}{C_1} + \frac{1}{C_2} \right) \right.$$

$$+ \frac{R}{LC_1} \left(\frac{qI_r}{KT} \right) \exp \left(\frac{qV_0}{KT} \right) \exp \left(\frac{qE \cos \omega t}{KT} \right) \bigg]$$

$$+ \frac{I_r(\beta + 1)}{LC_1 C_2} \exp \left(\frac{qV_0}{KT} \right) \exp \left(\frac{qE \cos \omega t}{KT} \right) = 0. \tag{I-16}$$

This equation can be expanded using the identity,

$$\exp Z \cos \theta = I_0(Z) + 2 \sum_{n=1}^{\infty} I_n(Z) \cos n\theta, \tag{I-17}$$

where $I_n(Z)$ represents modified Bessel function of the first kind and order n.

Retaining only the fundamental terms, after simplification, leads to the equation for oscillation shown below:

$$\omega^3 E \sin \omega t - \frac{R\omega^2 E}{L} \cos \omega t - \frac{\omega^2 VM}{C_1} \left[I_0(V) + I_2(V) \right] \cos \omega t \tag{I-18}$$

$$+ \frac{\omega^2 V^2 M}{2C_1} \left[I_1(V) - I_3(V) \right] \cos \omega t - \omega E \left[\frac{1}{L} \left(\frac{1}{C_1} + \frac{1}{C_2} \right) \right] \sin \omega t$$

$$- \frac{\omega RVM}{LC_1} \left[I_0(V) - I_2(V) \right] \sin \omega t + \frac{2(\beta + 1)MI_1(V) \cos \omega t}{LC_1 C_2} = 0.$$

Here use has been made of the trigonometric identities:

$$\sin^2 \theta = \tfrac{1}{2} - \tfrac{1}{2} \cos 2\theta$$

$$\cos^2 \theta = \tfrac{1}{2} + \tfrac{1}{2} \cos 2\theta$$

$$\cos x \cos y = \tfrac{1}{2} \cos (x + y) + \tfrac{1}{2} \cos (x - y)$$

$$\sin x \cos y = \tfrac{1}{2} \sin (x + y) + \tfrac{1}{2} \sin (x - y).$$

We have also defined $V = qE/KT$ and $M = I_r e^{qV_0/KT}$

In order for equation (I-18) to be satisfied, both the coefficients of the sine terms and the coefficients of the cosine terms must equate independently to zero. The equation resulting from the sine terms represents the frequency equation, and the equation from the cosine terms represents the amplitude equation.

We note also, using equations (I-2), (I-10), (I-12), and (I-17), that the emitter current is given by

$$I_e = (\beta + 1)I_r e^{(q/KT)(V + V_0)}$$

$$= (\beta + 1)M \left[I_0(V) + 2 \sum_{n=1}^{\infty} I_n(V) \cos n\omega t \right]. \tag{I-19}$$

The dc component is given by

$$I_e(\text{mean}) = (\beta + 1)MI_0(V) \qquad \text{(I-20)}$$

or

$$M = \frac{I_e(\text{mean})}{(\beta + 1)I_0(V)}. \qquad \text{(I-21)}$$

Turning now to the frequency equation resulting from the sine terms of equation (I-18), we have

$$\omega^2 = \frac{1}{L}\left(\frac{1}{C_1} + \frac{1}{C_2}\right) + \frac{RVM}{ELC_1}[I_0(V) - I_2(V)] = 0. \qquad \text{(I-22)}$$

Substituting for M from equation (I-21) and noting that $V/E = q/KT$, we have

$$\omega^2 = \frac{1}{L}\left(\frac{1}{C_1} + \frac{1}{C_2}\right) + \frac{RqI_e(\text{mean})}{KTLC_1(\beta + 1)}\left[\frac{I_0(V) - I_2(V)}{I_0(V)}\right] = 0. \qquad \text{(I-23)}$$

Using the identity $2I_1(V)/V = I_0(V) - I_2(V)$, and also noting from equation (G-27) that the base-to-emitter input resistance is given by

$$R_{\text{in}0} = \frac{(\beta + 1)KT}{qI_e(\text{mean})},$$

we have

$$\omega^2 = \frac{1}{L}\left[\frac{1}{C_1}\left(1 + \frac{2I_1(V)R}{VI_0(V)R_{\text{in}0}}\right) + \frac{1}{C_2}\right]. \qquad \text{(I-24)}$$

If we now define R_{in} by the relationship,

$$\frac{R_{\text{in}}}{R_{\text{in}0}} = \frac{VI_0(V)}{2I_1(V)}, \qquad \text{(I-25)}$$

equation (I-24) becomes

$$\omega^2 = \frac{1}{L}\left[\frac{1}{C_1}\left(1 + \frac{R}{R_{\text{in}}}\right) + \frac{1}{C_2}\right]. \qquad \text{(I-26)}$$

It is interesting to observe that equation (I-25) is the same as that derived in Appendix G, equation (G-28A), and that consequently the graph of Figure 3-6 can be used to determine R_{in} after the amplitude V has been determined from the amplitude equation (to be discussed subsequently).

Equation (I-26) cannot be conveniently used to determine the change in frequency resulting from a change in the amplitude of oscillation V because L, as we have used it, is the equivalent steady-state inductance of the crystal, which is itself a function of frequency. We may rewrite equation (I-26), however, in the form

$$\omega L = \frac{1}{\omega C_1}\left(1 + \frac{R}{R_{in}}\right) + \frac{1}{\omega C_2}. \tag{I-27}$$

Since the Q of the crystal is very high (normally several hundred thousand), the frequency of oscillation is always very near the resonant frequency of the crystal and the reactances of C_1 and C_2 may be considered to be constants the values of which are calculated at the nominal frequency of the crystal. We may then rewrite equation (I-27) as follows:

$$X_e + X_1\left(1 + \frac{R}{R_{in}}\right) + X_2 = 0, \tag{I-28}$$

where X_e is the crystal reactance and X_1 and X_2 are the capacitor reactances.

It is interesting to observe that the effect on frequency caused by R_{in} is the same as that which would be determined by placing the value of R_{in} determined from the nonlinear analysis of Appendix G into the linear Y-parameter, equation (7-12). This results from the $R_e X_1 g_{ie}$ term of K_2. Equation I-28 then becomes equivalent to equation (7-12) if we neglect the reactances of the transistor and the output conductance.

Turning now to the amplitude equation resulting from the coefficients of the cosine terms in equation (I-18), we have

$$-\frac{R\omega^2 E}{L} - \frac{\omega^2 VM}{C_1}[I_0(V) + I_2(V)]$$

$$+\frac{\omega^2 V^2 M}{2C_1}[I_1(V) - I_3(V)] + \frac{2(\beta+1)MI_1(V)}{LC_1C_2} = 0. \tag{I-29}$$

By making the substitutions, $E = VKT/q$ (as defined earlier) and

$$M = \frac{I_e(\text{mean})}{(\beta+1)I_0(V)}$$

[from equation (I-21)] and simplifying, we obtain the expression

$$R = \frac{qI_e(\text{mean})L}{2(\beta+1)KTC_1}\left[\frac{VI_1(V) - VI_3(V) - 2I_0(V) - 2I_2(V)}{I_0(V)}\right]$$

$$+\left(\frac{1}{\omega^2 C_1 C_2}\right)\left(\frac{qI_e(\text{mean})}{KT}\right)\frac{2I_1(V)}{VI_0(V)}. \tag{I-30}$$

Now using the identity

$$\frac{2n}{Z} I_n(Z) = I_{n-1}(Z) - I_{n+1}(Z),$$

(I-31)

with $n = 2$, we have

$$\frac{4I_2(Z)}{Z} = I_1(Z) - I_3(Z).$$

Making this substitution in (I-30) gives

$$R = \frac{-qI_e(\text{mean})L}{(\beta + 1)KTC_1} \left[\frac{I_0(V) - I_2(V)}{I_0(V)} \right]$$

$$+ \left(\frac{1}{\omega^2 C_1 C_2} \right) \left[\frac{qI_e(\text{mean})}{KT} \right] \frac{2I_1(V)}{VI_0(V)}.$$

(I-32)

Again using (I-31) for $n = 1$, we have

$$\frac{2I_1(Z)}{Z} = I_0(Z) - I_2(Z).$$

Performing this substitution gives

$$R = \frac{-qI_e(\text{mean})L}{(\beta + 1)KTC_1} \left[\frac{2I_1(V)}{VI_0(V)} \right]$$

$$+ \left(\frac{1}{\omega^2 C_1 C_2} \right) \left(\frac{qI_e(\text{mean})}{KT} \right) \frac{2I_1(V)}{VI_0(V)}.$$

(I-33)

Now let $R_{in0} = (\beta + 1)KT/qI_e(\text{mean})$ from equation (G-27) and

$$g_{m0} = \frac{qI_e(\text{mean})}{KT} \frac{\beta}{\beta + 1}$$

from equation (G-22). We have added the factor $\beta/(\beta + 1)$ here to account for the fact that $I_c = I_e\beta/(\beta + 1)$, since in Appendix G the assumption was made that $I_c \doteq I_e$. The resulting equation is

$$R = \frac{-L}{R_{in0}C_1} \left[\frac{2I_1(V)}{VI_0(V)} \right]$$

$$+ \left[\frac{1}{\omega^2 C_1 C_2} \right] \left[\left(1 + \frac{1}{\beta} \right) g_{m0} \right] \left[\frac{2I_1(V)}{VI_0(V)} \right].$$

(I-34)

Now observing that

$$\left(\frac{1}{\beta}\right)g_{mo} = \frac{qI_e(\text{mean})}{KT}\frac{\beta}{(\beta+1)}\left(\frac{1}{\beta}\right) = \frac{qI_e(\text{mean})}{KT(\beta+1)} = \frac{1}{R_{ino}},$$

and also substituting $X_1 = -1/\omega C_1$, $X_2 = -1/\omega C_2$, and $X_e = \omega L$, we have

$$X_1 X_2 g_{mo}\left[\frac{2I_1(V)}{VI_0(V)}\right] = R - X_1(X_e + X_2)\left[\frac{2I_1(V)}{VI_0(V)}\right]\frac{1}{R_{ino}}. \tag{I-35}$$

Now letting $g_m/g_{mo} = 2I_1(V)/VI_0(V)$ and $R_{in} = R_{ino} \ VI_0(V)/2I_1(V)$, as defined in Appendix G, so that Figure 3-6 can be used to determine the values, we have

$$X_1 X_2 g_m = R - \frac{X_1(X_e + X_2)}{R_{in}}. \tag{I-36}$$

It is interesting to note that this equation is consistent with linear equation (7-11) if we neglect the transistor reactances and the output admittance. Then K_1, equation (7-13), becomes $-X_1(X_2 + X_e)g_{ie}$, as we determined in equation (I-36) above.

Appendix J

Mathematical Development of the Sideband Level versus Phase Deviation Equation

The frequency spectrum of a phase or frequency modulated signal is well known and will not be derived here. A good treatment of the subject appears in several texts.[42]

The general form of the solution is

$$e = J_0(\delta) E_c \sin \omega_c t + J_1(\delta) E_c [\sin (\omega_c + \omega_m) t - \sin (\omega_c - \omega_m) t]$$

$$+ J_2(\delta) E_c [\sin (\omega_c + 2\omega_m) t + \sin (\omega_c + 2\omega_m) t] \cdots \quad \text{(J-1)}$$

where

e = resultant modulated signal,
E_c = peak unmodulated carrier voltage,
$J_n(\delta)$ = Bessel function of the first kind and of order n,
δ = deviation ratio (for frequency modulation, $\delta = f_d/f_m$; for phase modulation, δ is the peak phase deviation),
ω_c = carrier angular frequency,
ω_m = angular frequency of modulation $\omega_m = 2\pi f_m$, and
f_d = peak frequency deviation in hertz.

Bessel functions of the first kind are given by the infinite series,

$$J_n(\delta) = \frac{\delta^n}{2^n n!} \left[1 - \frac{\delta^2}{2(2n+2)} + \frac{\delta^4}{2(4)(2n+2)(2n+4)} \right.$$

$$\left. - \frac{\delta^6}{2(4)(6)(2n+2)(2n+4)(2n+6)} + \cdots \right]. \quad \text{(J-2)}$$

For small phase deviations ($\delta \ll 1$), only the carrier and first sideband pair are significant. The ratio of their amplitudes is given by $J_1(\delta)/J_0(\delta)$. From equation

(J-2) it can be seen that for small δ, $J_0(\delta) \doteq 1$ and $J_1(\delta) \doteq \delta/2$. The relative side-band level is then given by $\delta/2$, where δ is in radians.

The data for Figure 4-2 is given in decibels and degrees; therefore, it is neces-sary to modify the result, giving the equation

$$\frac{J_1(\delta)}{J_0(\delta)} = 20 \log \frac{\theta}{2(57.3)} \text{ dB} \tag{J-3}$$

where θ is in degrees.

Appendix K
Derivation of Crystal Equations

The equivalent circuit of a crystal is shown in Figure K-1. This circuit is well known, and the definitions of the components are as follows:

C_0 = holder capacitance,
L_1 = motional arm inductance,
C_1 = motional arm capacitance, and
R_1 = motion arm resistance.

If we define

$$Z_0 = \frac{-j}{2\pi f C_0} \tag{K-1}$$

and

$$Z_1 = R_1 + j\left(2\pi f L_1 - \frac{1}{2\pi f C_1}\right), \tag{K-2}$$

then the complex impedance of the crystal at any frequency is given by

$$Z_p = \frac{Z_0 Z_1}{Z_0 + Z_1} = \frac{\left[\dfrac{-j}{2\pi f C_0}\right]\left[R_1 + j\left(2\pi f L_1 - \dfrac{1}{2\pi f C_1}\right)\right]}{\left[R_1 + j\left(2\pi f L_1 - \dfrac{1}{2\pi f C_1} - \dfrac{1}{2\pi f C_0}\right)\right]} \tag{K-3}$$

$$Z_p = \frac{\left[\dfrac{2\pi f L_1 - (1/2\pi f C_1)}{2\pi f C_0} - \dfrac{jR_1}{2\pi f C_0}\right]}{\left[R_1 + j\left(2\pi f L_1 - \dfrac{1}{2\pi f C_1} - \dfrac{1}{2\pi f C_0}\right)\right]}. \tag{K-4}$$

For a resonance to occur, Z_p must be resistive and, therefore,

$$\frac{-R_1/2\pi f C_0}{\dfrac{2\pi f L_1 - (1/2\pi f C_1)}{2\pi f C_0}} = \frac{2\pi f L_1 - \dfrac{1}{2\pi f C_1} - \dfrac{1}{2\pi f C_0}}{R_1} \tag{K-5}$$

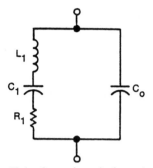

Figure K-1. Crystal equivalent circuit.

$$-R_1^2 = \left(2\pi f L_1 - \frac{1}{2\pi f C_1}\right)\left(2\pi f L_1 - \frac{1}{2\pi f C_1} - \frac{1}{2\pi f C_0}\right) \tag{K-6}$$

$$R_1^2 + 4\pi^2 f^2 L_1^2 - \frac{L_1}{C_1} - \frac{L_1}{C_0} - \frac{L_1}{C_1} + \frac{1}{4\pi^2 f^2 C_1^2} + \frac{1}{4\pi^2 f^2 C_1 C_0} = 0 \tag{K-7}$$

$$(2\pi f)^4 + (2\pi f)^2 \left(\frac{R_1^2}{L_1^2} - \frac{2}{L_1 C_1} - \frac{1}{L_1 C_0}\right) + \left(\frac{1}{L_1^2 C_1^2} + \frac{1}{L_1^2 C_1 C_0}\right) = 0. \tag{K-8}$$

Using the quadratic formula and solving for $(2\pi f)^2$ gives

$$(2\pi f)^2 = \frac{1}{2}\left\{\left(\frac{2}{L_1 C_1} + \frac{1}{L_1 C_0} - \frac{R_1^2}{L_1^2}\right)\right.$$
$$\left. \pm \left[\left(\frac{2}{L_1 C_1} + \frac{1}{L_1 C_0} - \frac{R_1^2}{L_1^2}\right)^2 - 4\left(\frac{1}{L_1^2 C_1^2} + \frac{1}{L_1^2 C_1 C_0}\right)\right]^{1/2}\right\}, \tag{K-9}$$

$$f = \frac{1}{2\pi}\left\{\left(\frac{1}{L_1 C_1} + \frac{1}{2L_1 C_0} - \frac{R_1^2}{2L_1^2}\right)\right.$$
$$\left. \pm \left[\left(\frac{1}{L_1 C_1} + \frac{1}{2L_1 C_0} - \frac{R_1^2}{2L_1^2}\right)^2 - \left(\frac{1}{L_1^2 C_1^2} + \frac{1}{L_1^2 C_1 C_0}\right)\right]^{1/2}\right\}^{1/2}. \tag{K-10}$$

We now consider the quantity,

$$\left[\left(\frac{1}{L_1 C_1} + \frac{1}{2L_1 C_0} - \frac{R_1^2}{2L_1^2}\right)^2 - \left(\frac{1}{L_1^2 C_1^2} + \frac{1}{L_1^2 C_1 C_0}\right)\right]^{1/2}$$

$$= \left(\frac{1}{L_1^2 C_1^2} + \frac{1}{L_1^2 C_1 C_0} - \frac{R_1^2}{L_1^3 C_1} + \frac{1}{4L_1^2 C_0^2} - \frac{R_1^2}{2L_1^3 C_0} + \frac{R_1^4}{4L_1^4} - \frac{1}{L_1^2 C_1^2} - \frac{1}{L_1^2 C_1 C_0}\right)^{1/2}$$

$$= \left[\left(\frac{1}{2L_1 C_0} - \frac{R_1^2}{2L_1^2}\right)^2 - \frac{R_1^2}{L_1^3 C_1}\right]^{1/2}. \tag{K-11}$$

For any practical crystal, however, it is normally true that

$$\left(\frac{1}{2L_1C_0} - \frac{R_1^2}{2L_1^2}\right)^2 \gg \frac{R_1^2}{L_1^3C_1}.$$

Then

$$\left[\left(\frac{1}{2L_1C_0} - \frac{R_1^2}{2L_1^2}\right)^2 - \frac{R_1^2}{L_1^3C_1}\right]^{1/2} \doteq \left(\frac{1}{2L_1C_0} - \frac{R_1^2}{2L_1^2}\right). \qquad \text{(K-12)}$$

Then

$$f \doteq \frac{1}{2\pi}\left[\left(\frac{1}{L_1C_1} + \frac{1}{2L_1C_0} - \frac{R_1^2}{2L_1^2}\right) \pm \left(\frac{1}{2L_1C_0} - \frac{R_1^2}{2L_1^2}\right)\right]^{1/2}. \qquad \text{(K-13)}$$

This equation gives two resonant frequencies; the first, obtained using the minus sign, is series resonance and the other is parallel resonance.

$$f_s = \frac{1}{2\pi}\left(\frac{1}{L_1C_1}\right)^{1/2} \qquad \text{(K-14)}$$

$$f_a = \frac{1}{2\pi}\left(\frac{1}{L_1C_1} + \frac{1}{L_1C_0} - \frac{R_1^2}{L_1^2}\right)^{1/2}. \qquad \text{(K-15)}$$

But

$$\frac{1}{L_1C_0} \gg \frac{R_1^2}{L_1^2}, \qquad f_a \doteq \frac{1}{2\pi}\left(\frac{1}{L_1C_1} + \frac{1}{L_1C_0}\right)^{1/2},$$

$$f_a \doteq \frac{1}{2\pi}\left(\frac{1}{L_1C_1}\right)^{1/2}\left(1 + \frac{C_1}{C_0}\right)^{1/2}. \qquad \text{(K-16)}$$

But $C_1/C_0 \ll 1$. Therefore, the binomial approximation,

$$(1 + x)^n \doteq 1 + nx \qquad \text{if} \qquad x \ll 1,$$

may be used, and

$$f_a = \frac{1}{2\pi}\left(\frac{1}{L_1C_1}\right)^{1/2}\left(1 + \frac{C_1}{2C_0}\right) = f_s\left(1 + \frac{C_1}{2C_0}\right). \qquad \text{(K-17)}$$

If $\Delta f = f_a - f_s$, then $\Delta f = (C_1/2C_0)f_s$ and the pullability $\Delta f/f_s = C_1/2C_0$.

The frequency at any load point C_L now can be calculated merely by making C_0 in the equation equal to the holder capacity plus the load capacity. Then

$$\frac{\Delta f}{f_s} = \frac{C_1}{2(C_0 + C_L)}. \qquad \text{(K-18)}$$

Substituting

$$\frac{C_0}{C_1} = r$$

gives

$$\frac{\Delta f}{f_s} = \frac{C_0}{2r(C_0 + C_L)}. \tag{K-19}$$

The resistance of the crystal at any load capacitance C_L can be found conveniently by redrawing the equivalent circuit of a crystal as given in Figure K-1. Here the resultant reactance of L_1 and C_1 is replaced by X, and the circuit of Figure K-2 results.

The impedance of the circuit may be written by inspection as

$$Z = \frac{(R_1 + jx)(jX_{C0})}{R_1 + j(X + X_{C0})}, \tag{K-20}$$

where $X_{C0} = -1/\omega C_0$.

Separating the real and imaginary parts gives, after some algebra,

$$Z = \frac{R_1 X_{C0}^2}{R_1^2 + (X + X_{C0})^2} + \frac{jX_{C0}[R_1^2 + X(X + X_{C0})]}{R_1^2 + (X + X_{C0})^2}. \tag{K-21}$$

The effective resistance is given by the real part and is

$$R_e = \frac{R_1 X_{C0}^2}{R_1^2 + (X + X_{C0})^2}. \tag{K-22}$$

To find the effective resistance at a specific load capacitance, it is necessary to determine the value of X at that load capacitance. This can be done with the aid of Figure K-3.

By definition, the crystal is operating at a load capacitance C_L when it is inductive and resonant with C_L. From Figure K-3 it can be seen that resonance of

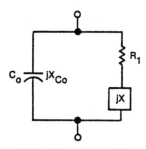

Figure K-2. Quartz crystal resonator.

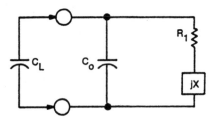

Figure K-3. Crystal and load capacitance.

the crystal with C_L occurs when resonance of X with $C_L + C_0$ occurs. Therefore when the crystal is operating into a load capacitance C_L the reactance X is given by

$$X = \frac{1}{\omega(C_L + C_0)}.$$

From equation (K-22), the effective resistance of the crystal is then given by

$$R_e = \frac{R_1 X_{C0}^2}{R_1^2 + \left[\dfrac{1}{\omega(C_L + C_0)} - \dfrac{1}{\omega C_0}\right]^2}. \tag{K-23}$$

This can be simplified to the form:

$$R_e = \frac{R_1 X_{C0}^2}{R_1^2 + \dfrac{1}{\omega^2 C_0^2}\left(-\dfrac{C_L}{C_L + C_0}\right)^2}$$

$$R_e = \frac{R_1 X_{C0}^2}{R_1^2 + X_{C0}^2\left(\dfrac{C_L}{C_L + C_0}\right)^2}. \tag{K-24}$$

If

$$\left|X_{C0}\left(\frac{C_L}{C_L + C_0}\right)\right| \gg R_1,$$

then

$$R_e \doteq \frac{R_1 X_{C0}^2}{X_{C0}^2\left(\dfrac{C_L}{C_L + C_0}\right)^2}$$

$$R_e = R_1\left(\frac{C_L + C_0}{C_L}\right)^2. \tag{K-25}$$

Appendix L
Sample Crystal Specification

DESCRIPTION

Metal-plated quartz resonator, wire-mounted in an HC-18/U holder; designed to operate on the fifth overtone mode of the fundamental frequency of the resonator under noncontrolled temperature conditions. The crystal unit shall be similar to type CR-56/U, per the latest version of MIL-C-3098, except for those paragraphs of this specification noted with a double asterisk (**) prefix.

1.	GENERATION INFORMATION
1.1.	Crystal element: AT-cut.
1.2.	Resonance: Series.
1.3.	Mode of vibration: Fifth overtone.
1.4.	Maximum drive level: 2.0 mW.
2.	ELECTRICAL PARAMETERS
2.1.	Frequency range: 50.0–125.0 MHz.
2.2.	Specified frequency: Attach table.
**2.3.	Frequency tolerance.
2.3.1.	Finishing (calibration) tolerance: ±0.0005 percent of the specified frequency when measured at +25°C (+1, -1°C).
2.3.2.	Drift tolerance (frequency stability): ±0.002 percent from the frequency measurement made at room ambient when measured over the operating temperature range.
2.3.2.1.	Method of measurement: Method B per the latest version of MIL-C-3098 and paragraphs 6.1 and 6.2 of this specification.
**2.4.	Test drive level: 1.0 ± 0.5 mW into 60 Ω.
**2.5.	Pin-to-pin capacitance: 4.5 pF, maximum.
2.5.1.	Method of measurement: Per latest version of MIL-C-3098 and paragraph 6.1 of this specification.
**2.6.	Equivalent resistance: 60 Ω, maximum, when measured over the operating temperature range.
2.6.1.	Method of measurement: Method B per latest version of MIL-C-3098 and paragraphs 6.1 and 6.2 of this specification.
**2.7.	Unwanted modes: Crystal units shall have a minimum unwanted

mode effective resistance of 120 Ω or an unwanted-to-main-mode resistance ratio of 3 to 1, whichever is greater.

3.   ENVIRONMENTAL REQUIREMENTS

3.1.   Operable temperature range: $-55°C$ $(+0, -3°C)$ to $+105°C$ $(+3, -0°C)$.

3.2.   Storage temperature range: $-65°C$ $(+0, -3°C)$ to $+105°C$ $(+3, -0°C)$.

3.3.   Shock: Crystal units shall meet the test requirements of paragraph 3.5 of this specification subsequent to testing in accordance with the latest version of MIL-STD-202, method 202, except for the details noted in the latest version of MIL-C-3098, method A.

**3.4.   Vibration: Crystal units shall meet the test requirements of paragraph 3.5 of this specification subsequent to testing in accordance with the latest version of MIL-STD-202, method 204, test condition A, except the limiting acceleration shall be 0.01 inch double amplitude or 5 g, whichever is less, and the details and exceptions noted in the latest version of MIL-C-3098.

3.5.   Vibration and shock test requirements: Maximum permitted change in frequency and equivalent resistance shall be as follows:
Permitted frequency change: ±0.0005 percent (±5 ppm).
Permitted equivalent-resistance change: ±10 percent.

3.6.   Leakage: In accordance with the latest version of MIL-C-3098.

3.7.   Insulation resistance: Crystal units shall have a minimum insulation resistance of 500 $M\Omega$ subsequent to testing in accordance with the latest version of MIL-STD-202, method 302, except for the details noted in the latest version of MIL-C-3098.

3.8.   Immersion: Crystal units shall meet the electrical requirements of paragraphs 3.7, 2.3, and 2.6 subsequent to testing in accordance with the latest version of MIL-STD-202, method 104, test condition B.

3.9.   Salt spray: Crystal units shall show no visible evidence of corrosion in addition to meeting the electrical requirements of paragraphs 3.7, 2.3, and 2.6 subsequent to testing in accordance with the latest version of MIL-STD-202, method 101, test condition B.

3.10.   Moisture resistance: Crystal units shall meet the electrical requirements of paragraphs 3.7, 2.3, and 2.6 subsequent to testing in accordance with the latest version of MIL-STD-202, method 106, except for the details and exceptions noted in the latest version of MIL-C-3098.

3.11.   Aging: The permitted change in frequency from the highest to the lowest measurements shall be 0.0005 percent (5 ppm) when tested in accordance with the latest version of MIL-C-3098 at $+85°C$ $(\pm2°C)$ for a period of 30 days.

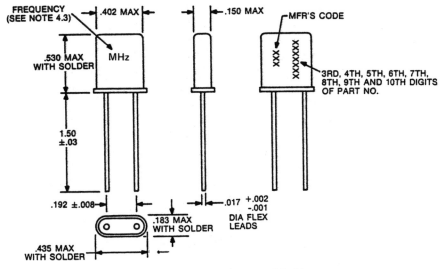

FREQUENCY (SEE NOTE 4.3)

.402 MAX

.150 MAX

MFR'S CODE

.530 MAX WITH SOLDER

MHz

XXX XXXXXX

3RD, 4TH, 5TH, 6TH, 7TH, 8TH, 9TH AND 10TH DIGITS OF PART NO.

1.50 ±.03

.192 ±.008

.017 +.002 -.001 DIA FLEX LEADS

.183 MAX WITH SOLDER

.435 MAX WITH SOLDER

Figure L-1. HC-18/U crystal holder.

3.12. Low-temperature storage: Per latest version of MIL-C-3098.

4. MECHANICAL REQUIREMENTS

4.1. Holder: HC-18/U with leads tinned suitable for soldering. (See Figure L-1.)

**4.2. Markings: Shall include frequency in megahertz, purchaser's part number, manufacturer's code, and date of manufacture. Markings shall be in accordance with the latest version of MIL-C-3098.

4.3. Example crystal frequency markings:

Below 100 MHz . . . . XX.XXXXX MHz.

100 MHz and above . . . . XXX.XXXX MHz.

5. QUALITY ASSURANCE PROVISIONS

5.1. Receiving inspection: Each lot of material submitted to this specification will be inspected in accordance with a sampling plan approved by quality control and quality assurance departments. Acceptance of the lot will be determined upon successful measurement of the following critical and/or major characteristics:

    a. Visual and mechanical inspection (external) . . . (major).

    b. Frequency at room ambient and over the operating temperature range . . . (major).

    c. Equivalent resistance at room ambient and over the operating temperature range . . . (major).

    d. Pin-to-pin capacitance at room ambient . . . (major).

     e. Unwanted modes at room ambient . . . (major).

     f. Leakage, in accordance with the latest version of MIL-C-3098 . . . (major).

5.2.   Quality control: Sample quantities from parts furnished on production orders may be tested to any requirement specified herein and disassembled to check for quality of workmanship by the purchaser.

5.3.   Design change approval: Any deviations in the manufacturing process or materials used in preparing the component evaluation and/or engineering samples must be approved by the company's engineering division.

6.   TESTING

6.1.   Method of testing: The crystal unit holder shall be ungrounded when making frequency, pin-to-pin capacitance, and equivalent resistance measurement. The lead length for test shall be $\frac{1}{4}$ inch ($\pm\frac{1}{16}$ inch) from the holder base.

6.2.   Test equipment: TS-683/TSM with a Radio Frequency Lab model HB8770 adaptor.

6.3.   Frequency correlation: The frequency of a given crystal unit, as measured by the supplier's equivalent test set, shall agree with or be within ±0.0005 percent (±5 ppm) of the same measurement made with the Government reference standard test set, or the company's equivalent test set.

6.4.   Resistance correlation: The equivalent resistance of a given crystal unit, as measured by the supplier's equivalent test set, shall agree with or be within ±10 percent of the same measurement made with the Government reference standard test set or the company's equivalent test set.

6.5.   Government source inspection marking: When parts to be supplied are required to have government source inspection as indicated by the part number shown on the purchase order, said parts shall be identified as Government-inspected items by the letter "G" stamped in ink or paint on each item prior to shipment. The letter "G" shall be of sufficient size to be easily identified by a person with normal or corrected vision, but not so large that it interferes with other markings. Method of marking is optional but must be capable of withstanding abrasion and scuffing that may be encountered in normal handling and shipping.

Bibliography

1. Anderson, A. E., Frerking, M. E., and Hykes, G. R., "Temperature-Compensated Quartz Crystal Units." Report No. 10, Contract No. DA-28-043-AMC-00210(E), Department of the Army, Project No. 1G622001A05801, April 1965.
2. Armour Research Foundation, "A Study of Crystal Oscillator Circuits." Final Report, Contract No. DA-36-039-sc-64609, August 1957, p. 37.
3. Battelle Memorial Institute, "Crystal Ovens." Contract No. DA-36-039-sc-73212, ASTIA Document No. AD210095, September 1958.
4. Bechmann, R., "Frequency–Temperature–Angle Characteristics of AT-Type Resonators Made of Natural and Synthetic Quartz." *Proc. IRE*, Vol. 44, No. 11, November 1956, pp. 1600–1607.
5. Bechmann, R., "Frequency–Temperature–Angle Characteristics of AT- and BT-Type Quartz Oscillators in an Extended Temperature Range," *Proc. IRE*, Vol. 48, No. 8, August 1960, p. 1494.
6. Buchanan, John P., *Handbook of Piezoelectric Crystals for Radio Equipment Designers.* Philco Corporation, WADC Technical Report 56-156, ASTIA Document No. AD110448.
7. Buroker, G. E. and Frerking, M. E., "A Digitally Compensated TCXO." *Proc. 27th Annu. Frequency Control Symposium*, 1973. (Also see U.S. Patent No. 3713033 by Frerking).
8. Chi, A. A., "Frequency–Temperature Behavior of AT-Cut Quartz Resonators." *Proc. 10th Annu. Symposium Frequency Control*.
9. Chow, W. F., "Crystal Controlled High Frequency Oscillators." *Semiconductor Products*, September 1959.
10. Cunningham, W. J., *Introduction to Nonlinear Analysis.* New York: McGraw-Hill Book Company, 1958.
11. Diamond Ordnance Fuze Laboratories, "Variation of Complex Y-Parameters with Emitter Current and Temperature." ASTIA Document No. AD245667, October 1960.
12. Frerking, M. E., "Short Term Frequency Stability Measurements." *Proc. 21st Annu. Symposium Frequency Control*, April 1967, pp. 273–286.
13. Frerking, M. E., "Spurious Oscillation in Crystal Oscillators," *Proc. 20th Annu. Symposium Frequency Control*, April 1966, pp. 501–516.
14. Frerking, M. E., "Vector Voltmeter Crystal Measurement System." *Proc. 23rd Annu. Symposium Frequency Control*, May 1969, pp. 93–101.
15. Georgia Institute of Technology, "Transistor Oscillators of Extended Frequency Range." Final Report, Contract No. DA-36-039-sc-42712, ASTIA Document No. AD82523, June 1955.
16. Georgia Institute of Technology, "VHF and UHF Crystal Controlled Oscillators." Quarterly and Final Reports, Contract No. DA-36-039-sc-85048, ASTIA Documents Nos. 247494, 253394, 267277, June 1961.
17. Gerber, E. A., "Reduction of Frequency Temperature Shift of Piezoelectric Crystals by Application of Temperature-Dependent Pressure." *Proc. IRE*, Vol. 48, No. 2, February 1960, pp. 244 and 245.
18. Hafner, E., "Crystal Resonators." *IEEE Trans. Sonics Ultrason.*, Vol. SU-21, No. 4, October 1974.

19. Hafner, E., "The Piezoelectric Crystal Unit–Definitions and Methods of Measurement." *Proc. IEEE*, February 1969, pp. 179–201.

20. Hellerman, H., *Digital Computer System Principles.* New York: McGraw-Hill Book Company, 1973.

21. Holford, K., "Transistor *LC* Oscillator Circuits for Low-Frequency, Low-Power Operation." *Mullard Tech. Commun.*, No. 41, December 1959.

22. Holford, K., "Transistor *LC* Oscillator Circuits Giving Modest Values of Power." *Mullard Tech. Commun.*, No. 42, February, 1960.

23. Hykes, G. R. and Newell, D. E., "A Temperature-Compensated Frequency Standard," *Proc. 15th Annu. Frequency Control Symposium*, pp. 297–317.

24. Intel Corporation, "Intel 8080 Microcomputer Systems User's Manual." Santa Clara, California, September 1975.

25. "IRE Standards on Piezoelectric Crystals–The Piezoelectric Vibrator: Definitions and Methods of Measurement." *Proc. IRE*, Vol. 45, No. 3, March 1957.

26. Johnson, George, "Power Gain and Stability in Linear Active Two-Ports." Texas Instruments RF seminar papers.

27. Koerner, L. F., "Methods of Reducing Frequency Variation in Crystals Over a Wide Temperature Range." *1956 IRE Conv. Record*, Part 8, pp. 48–54.

28. Magnavox Company, "Quartz Crystal Oscillator Circuits." First Quarterly Progress Report, Contract No. DA-36-039-sc-88892, ASTIA Document No. AD269990, September 1961.

29. Magnavox Company, "Quartz Crystal Oscillator Circuits." Second Quarterly Progress Report, Contract No. DA-36-039-sc-88892, ASTIA Document No. AD273873, December 1961.

30. Magnavox Company, "Quartz Crystal Oscillator Circuits," Third Quarterly Progress Report, Contract No. DA-36-039-sc-88892, ASTIA Document No. AD278234, March 1962.

31. Magnavox Company, "Quartz Crystal Oscillator Circuits." Final Report, Contract No. DA-36-039-sc-88892, ASTIA Document No. AD297361, June 1962.

32. Magnavox Company, "Quartz Crystal Oscillator Circuits Study," Report No. 1, Contract No. DA-36-039-AMC-00043(E), April 15, 1963.

33. McLachlan, N. W., *Bessel Functions For Engineers*, 1st Edition. Amen House, London: Oxford University Press, 1934.

34. MIL-C-3098F, 28 Nov 1972.

35. Motorola, Inc., "Transistor Crystal Oscillator Circuitry." Final Report, Contract No. DA-36-039-sc-72837, August 1957.

36. Motorola Semiconductor Products, Inc., "M6800 Microprocessor Programming Manual." Phoenix, Arizona, 1975.

37. Mroch, Alan, and Hykes, Glen, "High Stability Temperature Compensated Crystal Oscillator Study." Third Interim Report, Contract No. DAAB07-73-C-0137, Fort Monmouth, N.J., U.S. Army Electronics Command, February 1975.

38. NavShips 93484, "Selected Semiconductor Circuits." No. 73231, Transistor Applications, Inc.

39. Pritchard, R. L., "Modern High-Frequency Transistors." Paper presented at the 1957 IRE–AIEE University of Pennsylvania Transistor and Solid-State Circuits Conf.

40. Reddi, G., "Application of High Frequency *Y*-Parameters." APP-3, Mountain View, California, Fairchild Applications Data, December 1958.

41. Reich, Herbert J., *Functional Circuits and Oscillators.* Princeton, New Jersey: D. Van Nostrand Company, Inc., 1961.

42. Seely, Samuel, *Electron-Tube Circuits.* New York: McGraw-Hill Book Company, Inc., 1958, pp. 600–610.

43. Seshu, Sundaram and Balabanian, Norman, *Linear Network Analysis.* New York: John Wiley and Sons, Inc., 1959.

44. Signal Corps, "Fundamental Principles of Crystal Oscillator Design." ASTIA No. AD39664, Long Branch Signal Laboratory, U.S. Army, Circuit Section, 1945.

45. Texas Instruments Co., Engineering Staff, *Transistor Circuit Design.* New York: McGraw-Hill Company, Inc., p. 312.

46. Vitro Laboratories, "Reliable Preferred Solid-State Functional Divisions." Contract No. AF30(602)1906, ASTIA Document No. 254184, December 1959.

47. Peduto, Ralph, and Prak, Jan Willem L., "Digital IC's Set Temperature Compensation for Oscillators." *Electronics*, August 14, 1972, pp. 124–126. (Also see U.S. Patent No. 3719838 by Peduto and Prak.)

48. Wilson, G. H., "Nonlinear Analysis of Sinusoidal Oscillators," ASTIA Document No. AD272912, University of California, Electronics Research Laboratory, December 1961.

49. Basic Method for the Measurement of Resonant Frequency and Equivalent Series Resistance of Quartz Crystal Units by Zero Phase Techniques in a π-Network, IEC Publication 444 (1973).

50. Proc. 30th Annu. Symposium Frequency Control, 1976, Sponsored by U.S. Army Electronics Command, Fort Monmouth, N.J., Published by Electronic Industries Association, 2001 I Street N.W., Washington, D.C. 20006.

51. Peregrino, L. and Ricci, D., "Phase Noise Measurement Using a High Resolution Counter With On-Line Data Processing." *Proc. 30th Annu. Symposium Frequency Control*, 1976, pp. 309–317.

52. Barns, J. *et al.*, "Characterization of Frequency Stability." *IEEE Trans. Instrum. Meas.*, Vol. IM-20, No. 2, May 1971.

53. MIL-STD-683E, 26 Aug 1975.

54. Blair, B. E., *Time and Frequency: Theory and Fundamentals*," NBS Monograph 140, 1974, US Government Printing Office, Washington, D.C.

55. Ballato, A., "Temperature Compensated Crystal Oscillator (TCXO) Design Aids: Frequency–Temperature Resonator Characteristics as Shifted by Series Capacitors," Electronic Technology and Devices Laboratoy, ECOM.

56. Hellwig, H., Frequency Standards and Clocks: A Tutorial Introduction, NBS Technical Note 616, U.S. Department of Commerce/National Bureau of Standards, 1977.

57. Buroker, G. E., High-Stability Temperature-Compensated Crystal Oscillator Study (HSTCXO), Final Report, Contract DAAB07-71-C-0136, United States Army Electronics Command.

58. Hafner, E., Analysis and Design of Crystal Oscillators, U.S. Army Electronics Laboratories Technical Report ECOM-2474, May 1964.

59. Bottom, Virgil E., *The Theory and Design of Quartz Crystal Units.* McMurry Press, 1968.

60. Heising, Raymond A., *Quartz Crystals for Electrical Circuits.* New York: D. Van Nostrand Company, Inc., 1946. Reprinted 1978, Electronic Industries Association, Washington, D.C.

Index

Charity was laughing at her now, after slapping the purple flower from her hand.

"You were trying to fit it into your hair, the way Evan Dobson did with that flower at the dance in Nebraska, weren't you?" More sharp laughter. "Foolish girl. Don't you realize he pitied you? That's the only reason he asked *you* to dance instead of *me*. He knew we were leaving the next day. He felt sorry for you."

"He told me I was beautiful," Faith said quietly, trying hard to believe he meant it, for nobody had ever told her that before.

Charity's words drowned her attempt in the flood of her chastening arguments. "You are far from beautiful," she said abruptly. "I guess it's up to me to tell you the truth, however much it hurts. I'm only doing it for your own good."

Was it true? Did he feel sorry for her? Because she was blind? Was she indeed ugly? Blind and ugly—how could God have afflicted her so? She would probably never marry; never bear children of her own; never know the wonders of being a woman. Why? What was her purpose then? A cloud of depression swarmed over her, stinging like a nest of angry wasps.

KELLY R. STEVENS is a multi-published, award-winning author of historical and contemporary novels. She spent a portion of her career as a journalist, while raising her children, and has had numerous features published. A feature Kelly wrote on some local history made its way into *True West Magazine,* a national publication. Kelly lives in the Texas Panhandle with her artist husband and her daughter.

HEARTSONG PRESENTS

Books by Kelly R. Stevens
HP106—Ragdoll
HP304—Song of the Cimarron

One with
the Wind

Kelly R. Stevens

Heartsong Presents

This book is dedicated to my husband, Grant Johnson, the soul mate I had been searching for all my life. I thank God that we are together in love, matrimony, creative ideals, and, most of all, our unified belief in our Lord and Savior, Jesus Christ.

I would like to thank the following people for their help and encouragement during the writing of this book: Doris and Virgil Varner for their research efforts; Dwaine Nolte and Jean Huckins for reading portions of the final draft to make sure I had my concepts on various issues worded for clarity and conciseness; Kelleen and Miah Ebel, Violet and Orval Johnson, Stacey Johnson, Steve and Sally Varner, Robin Jones, Zella Giuliano, and all of the above for their undying encouragement toward my efforts. And, most of all, I would like to thank my husband, Grant Johnson, for his support, encouragement, gentle prodding, and, above all, love, as I created a work I believe in with all my heart.

A note from the author:
I love to hear from my readers! You may correspond with me by writing: **Kelly R. Stevens**
Author Relations
PO Box 719
Uhrichsville, OH 44683

ISBN 1-58660-329-9

ONE WITH THE WIND

Cover illustration by Jocelyne Bouchard.

one

Colorado, 1870

"Make sure the groove is wide enough down yer side of the chimney," Grady Lambert told his son, "so the logs'll fit."

They had labored long hours under the hot Colorado sun building their home near the Rio Grande River, fourteen miles east of Del Norte. This was the closest settlement and a popular gathering place for freighters hauling supplies to the mines in the San Juan Region. From here the Lamberts could easily sell farm goods in Del Norte.

Virgil studied the groove and lifted his dingy, white Stetson off his head, letting the light breeze cool the sweat that had gathered. He wiped his forehead and replaced the hat.

"I am," he answered. "Same size as the other."

The nineteen year old had never worked so hard in his life. The June sun beat down on his back, creating a river of moisture along his spine. He was glad the chimney stood erect, the doors had been completed, and the sill logs were in place. Once the logs were cut and notched, they could then be positioned one on top of the other to create the walls for the home.

Peering down on his two brothers, Joshua and James, Virgil watched them work, giving himself enough time to catch his breath.

Virgil admired his brother James, three years younger, for his dark good looks and his way with the ladies. James and their sister Charity took more after their mother, while the rest of the Lambert siblings resembled their father.

Joshua, his fraternal twin, was larger than he, although they shared the same sandy blond hair and gray eyes. Unfortunately, Joshua was born simple-minded, and Virgil couldn't overcome the guilt he felt about this. He believed he might be partly responsible, since they had occupied the same womb at the same time. The delivery had been difficult for their mother. Virgil was delivered first, and Joshua, ten minutes later. Two healthy babies had entered the world, or so they believed. Early on, they discovered something wasn't quite right about Joshua.

Virgil watched over his twin like a shepherd guarding sheep; the boy was often stumbling or bumping into something. The children back in Nebraska had teased Joshua until he cried. Poor Joshua didn't know any better. He tried to be a part of things and fit in. Sometimes Virgil wondered if his brother even knew he was different.

A smile stretched across his face when he thought of his sister Faith. She *knew* she was different. Glancing her way where she sat in a grassy meadow surrounded by wild flowers, he noticed she was struggling to fit something into her hair. He knew she was remembering that Dobson boy, and his smile melted into a frown.

Then he caught his father's stern gaze. Quickly Virgil set back to work on the log home.

Faith had, indeed, been remembering Evan Dobson.

But her own smile changed into a frown as well when she heard the commotion nearby.

Her sister, Charity, eighteen, was apparently trying to milk the cow—to no avail.

"Still no milk! Dumb beast!"

Faith cringed, knowing her sister was in one of her "moods" again. She heard her yell in their mother's direction.

"Ma! She ain't gonna give!"

The clank of the pail as it hit the side of the schooner made her flinch. Faith continued to listen, though wishing to be left to her memories in her secret world of perfection.

"She's skin and bones," Eliza Lambert explained. "It'll take time for 'er to fill out again. Jest need to keep workin' 'er."

Faith breathed deeply as she caught the scent of rabbit stew. She could hear the muted sound of metal against metal as her mother stirred the stew in the pot hanging from a tripod over the fire. That pleasure was also short lived when she heard her mother's next words.

"Go check on Faith for me."

No! Faith wanted to yell. Now she would have to endure Charity's wrath, and she was growing weary of it.

"I'm tired of baby-sitting her!"

Faith's heart skipped a beat at her sister's words. Next she heard what sounded like a rock flying and striking the ground some distance away and knew Charity had kicked the stone in anger.

"She's twenty-one years old now, Ma. She can take care of herself!"

"Do as yer ma says," warned Mr. Lambert, passing by on his way to check the notches James had made in the cottonwood logs.

Some time passed, and Faith thought she had been forgotten. She relaxed and delved into her memories again. She fought the lupine for several minutes, but it wouldn't stay fixed in her hair. Suddenly, and to her dismay, she heard footsteps approach. Her only recourse was to pretend nothing had happened.

"Hi, Charity," she called out cheerfully, knowing it was her younger sister.

Charity had a strange way of walking, almost a shuffle as she moved from place to place. Of course, the girl would never believe it. Joshua's steps, as well, were much different from Virgil's or James's. Joshua walked heavily, as if he carried a great burden on his shoulders and it weighed him down.

"Did she give milk?" Faith asked, trying to make polite conversation as she continued to fumble with the now limp flower. It had been several days since their cow had ceased providing milk for the family. For a second Charity was silent.

"No," she finally replied. "How did you know I tried to milk the cow?"

Tilting her face toward Charity's voice, Faith responded, "I smell cow on you."

"No, you can't!" Charity bellowed. "Don't you dare say I smell like cow!"

Faith tensed. Why was her sister angry all the time? She didn't mean to hurt her with the comment, but it was the truth. Charity regularly responded this way no matter what she said or did. Her "moods" were becoming more frequent.

"I'm sorry," Faith said. "You always smell real nice to me," she added, hoping the strain in her sister's voice would soften.

Charity had a wonderful, natural scent, unlike the boys. Sometimes they could smell pretty bad, especially after working in the sun all day. Yet they each had their own scents under normal conditions, as did every person with whom she came in contact. Some smelled nice, while others were plain awful.

She remembered the Indian in Nebraska who was always drunk on sour mash. It wasn't that he'd smelled so bad, but he smelled wild, like the outdoors, like horses too. The scent of liquor detracted from the pleasantness of the other odors.

"What's that you're holding?" Charity asked, changing the subject.

Faith held the droopy plant out in the palm of her hand. Every time she smelled or touched a flower, it rekindled memories of a kind young man, though she shared this with no one. "Is this a flower?" she asked, knowing it was. "Is it pretty?"

A hard snicker escaped Charity's lips. "A dead flower," she said sharply. "You've worn it out with all your touching. What are you doing with it anyhow?"

Faith felt the smack against her hand as Charity slapped the lupine from her fingers. Another snicker mingled with the air when Faith shrunk back.

"I–I was trying to put it in my hair," Faith explained, suddenly fearful.

One time Charity had slapped her and hard. Faith placed both palms on her cheeks and could still feel the sting of the blow, as though it had just happened. The memory made her tremble.

Relatives living in Colorado had succeeded, through letters, in convincing Grady and Eliza Lambert to make the trek to the San Luis Valley. Mr. Lambert was a farmer in Prague, Nebraska, and the relatives had assured him Colorado soils could grow anything better and faster. "God's country," they called it. A section of land near the Rio Grande River was for sale. The previous owners had made a hurried departure unexpectedly.

Rumors told how the original settler, a Mr. Rucker, had killed a Cheyenne warrior who was simply passing through with his companion, another warrior, on his way to trade with the Utes. Rucker thought the Indian was up to no good and shot him in the back. His comrade escaped, unharmed. Less than twenty-four hours later, Rucker packed up his family and escaped to "safer ground," as he described it.

Neighbors who lived nearby had commented on how James, though much younger, resembled Mr. Rucker in stature and coloring. Concern mounted, but the Lamberts were assured of complete safety, with Fort Garland being so close to their settlement, less than forty miles away. Now the Lamberts owned the land, having arrived a little over a month ago.

The night before they had left Nebraska to begin the long, arduous journey to the Southwest, Eliza Lambert had let the children attend a barn dance. It would be the last opportunity to say good-bye to friends and acquaintances.

Faith loved the music and clapped her hands to the beat. All Charity did was complain, until she caught the eye of a familiar young man.

"There's someone over there who keeps looking at me," she whispered to Faith. "Wait! I've seen him before. It's Evan Dobson, the kid we went to school with—except he's grown up, changed. He's about our age, maybe a little older." She hesitated. "You should see him, Faith. He's got light streaks in his brown hair, and he's dark and muscular. From working in the sun, I guess—I heard he's a logger. Oh, and he's got the most wonderful green eyes. Mercy!"

"Will you dance with him if he asks you?"

Faith didn't understand the activity, but she had listened to the shuffling of feet moving across the floor. The steps seemed to become a part of the music itself. She knew it was a "coupled" thing between a man and a woman, a boy and a girl. Faith liked the wind that came off the dancers as they passed, brushing across her cheeks and blowing tendrils of hair off her shoulders. Their laughter kept a smile on her lips throughout the evening.

"Of course, I'll dance with him, you ninny! But who will dance with you?" The remark was full of spite.

Faith felt her fingernails dig into the palms of her hands. "I–I don't know how," she answered, trying to keep tears at bay.

"Nobody would even *ask* you," Charity said abruptly.

Faith said nothing. Arms to her sides, she grasped a section of her faded calico between the fingers of one hand. Heat flooded her face, and she tried to cool the sensation by securing her palms on each cheek. She hated this emotion, and it always seemed Charity could bring it bubbling to the surface whenever she chose. She later learned Charity had given her the most faded dress they owned, while she took the brightest gingham for herself.

The green material matched Charity's eyes, as she'd hoped. It didn't matter that the faded blue of Faith's dress matched her own ice blue eyes with stunning effects. Charity didn't think anybody would notice. Further, her neckline was swathed in creamy lace, while Faith's neckline was buttoned plainly against her throat.

Charity had gone to great lengths to braid her own thick hair and roll it around her head in a dazzling array. Dark ringlets dropped like springy coils about her face and neck. She had combed Faith's hair, as her mother asked, but left it hanging loose over the shoulders with no style to enhance her blind sister's features.

"Look!" Charity said in hushed tones. "He's coming my way."

Faith brightened. She was happy for her sister. Besides, if Charity was content, she generally relaxed her attacks, which had seemed to increase with the plans to leave their homeland. Why, Faith didn't know.

Then Charity gasped. "Oh—wait!"

"What? Is he coming?"

For a second Charity didn't respond. "He started to, then he just up and went out the door." Another gasp. "Wait! He's back now."

Suddenly, she felt Charity grasp her arm tightly with one hand. "Is he coming?" Faith repeated, each word filled with anticipation.

Tightening her grip on Faith's arm, Charity whispered,

"Keep your voice down. I don't want him to know I'm watching.

"Oh," she continued from one side of her mouth, "he's carrying a flower—a daisy, I think." Charity paused for breath. "That's why he went outside! He's bringing me a flower, in exchange for a dance. How wonderful!"

In times such as these Faith would give anything for even a brief moment of sight, that miraculous something God had chosen to keep from her. Why? she wondered. If only she could see her sister's face right now, it was probably wrapped in delight. She heard the joy in Charity's voice. Charity loosened her hold on Faith's arm. All senses piqued as Faith prepared for the glorious moment.

"He's so handsome," were her sister's last whispered words before she fell completely silent.

Faith studied his tone when he spoke the awaited question.

"May I have this dance?"

The voice seemed bottomless, resonating from somewhere deep within him, yet smooth, tranquil. It filled Faith with instant peace. His breath smelled of cinnamon as its breeze touched her nose. Faith marveled at how his sound and fragrance calmed her racing heart. Yes, he had changed.

She sensed the warmth from his body, and it made the hairs on her arms rise. His presence revealed his countenance to her in a most personal way. To Faith, he *felt* like a good person, a confident person. Charity's next words sent a chill through her veins.

"She's blind, you ninny. Can't you see that?"

"I know she's blind," came the smooth reply. "What does that matter?"

Faith felt sudden pressure against her temple and jerked away from the touch.

"It's okay," said Evan Dobson, assurance in his voice. "I only want to put this daisy behind your ear. It's gold," he added, "the same color as your hair."

Even though Faith heard her sister expel a charged breath, she leaned toward the voice, allowing him to work the stem of the flower into her long hair before securing it behind her right ear.

"There," he told her. "I wish you could see yourself. You're the most beautiful woman here."

"I told you she's blind!" Charity stamped her foot. "And she can't dance either, so you're wasting your time."

"Will you try?" Evan asked pleasantly, ignoring Charity's words. "You only have tonight. Tomorrow you'll be gone."

A few families planned to travel to Colorado, his own included, although the Dobsons wouldn't leave until next spring.

Faith's heart beat hard and fast. *What should I do? I've never danced before—never even been asked!* Yet somehow she trusted this person whose confidence rose up around her like some comforting force. And what had he called her? Beautiful?

"Go ahead," a voice spoke softly from behind. It was her mother, who left as quickly as she'd arrived.

Evan tried again. "I'll be careful with you—I promise."

At this Faith stepped forward, toward his sound, losing herself in it. She lifted her face and spoke. "I'll try, if you'll help me."

"Thatta girl," Evan told her.

Faith felt the gentle pressure applied to her wrists and

the warmth that enfolded them as he circled strong fingers around both. "What do I do?"

"I'll lead." Evan took her hands gently. "Just follow me. I promise I won't let anybody bump into you."

A giggle rose in Faith's throat and found its way out between her lips. "It isn't the others I'm worried about," she warned. "I'm afraid *I'll* bump into *them.*"

"You're doing fine," Evan said, with a slight chuckle of his own. "I love the way you laugh. I've never heard a laugh like that before. It's beautiful, delicate, yet full at the same time." He paused and led her to the outside edge of the dance floor. "And so genuine. I can tell a lot about a person by the way they laugh."

That required more explanation, but Faith couldn't dwell on the words. She could only deal with one thing at a time. Right now it was the hot sensation that filled her cheeks. The feeling wasn't bad; in fact, she liked it very much.

But just then Charity stepped up and took hold of Faith's arm in a viselike grip.

"Ouch!"

"It's time for you to go to the wagon!" Charity said, hot anger in her voice. "You're behaving like a trollop!"

"Oh, no, I didn't mean—"

"Leave her be," Evan said, a low growl in his voice.

"No!" Faith said quickly. "I must go! I never meant—"

Evan's voice grew faint as Faith was led away into the night. Why had she allowed such a thing? Faith wondered. How could this have happened? Tears rolled down her cheeks as she was shoved to the side of the family wagon. She nearly stumbled but used the spoked wheel to right herself.

"Ma and Pa will surely hear of this!" Charity blurted out. "You should have seen yourself! Primping like a Jezebel!"

"I didn't mean—" Suddenly, anger welled up in her chest. "All I was doing was following him! I couldn't have been 'primping,' as you say, because I was too busy concentrating on stepping. I was only worrying about who I might run into!"

Charity let out her breath. "It wasn't just how you were moving," she said haughtily. "It's how you're dressed and that ridiculous flower in your hair!"

"You were pleased when he brought the flower, thinking it was for you!" Faith's body trembled; her teeth were clenched. "You would have taken it, and you know it! As far as the way I'm dressed," Faith continued, her hand gripping the collar at her throat, "you're the one who dressed me! So if I look like a Jezebel, you did it!" She nearly screamed the last words. It was then she felt the blow.

Her left cheek stung, and she fell back against the wagon again, this time slumping to the ground in a heap.

"Don't you ever use that tone of voice with me again, you hear?" Charity said with contempt in her voice. "Or Mother and Father *will* hear of this!"

Faith swung her arms out, trying to grab at her sister's dress, something, anything. She could grasp nothing but cool, night air. Charity had no right to hit her or scold her. And the laughter. Charity mocked her now. Faith knew she shouldn't strike back, should turn the other cheek, but—

"Poor little blind mouse," the younger girl said scornfully. "What do you think you could accomplish if you did get hold of me? Huh?"

Hot tears burned Faith's eyes as she tried to force out the sound of her sister's laughter. But it seemed to grow in intensity, filling her head, making it ache.

Suddenly she was brought back from the memory. Charity was laughing at her now, after slapping the purple flower from her hand.

"You were trying to fit it into your hair, the way Evan Dobson did with that flower at the dance in Nebraska, weren't you?" More sharp laughter. "Foolish girl. Don't you realize he pitied you? That's the only reason he asked *you* to dance instead of *me*. He knew we were leaving the next day. He felt sorry for you."

"He told me I was beautiful," Faith said quietly, trying hard to believe he meant it, for nobody had ever told her that before.

Charity's words drowned her attempt in the flood of her chastening arguments. "You are far from beautiful," she said abruptly. "I guess it's up to me to tell you the truth, however much it hurts. I'm only doing it for your own good."

Was it true? Did he feel sorry for her? Because she was blind? Was she indeed ugly? Blind and ugly—how could God have afflicted her so? She would probably never marry, never bear children of her own, never know the wonders of being a woman. Why? What was her purpose then? A cloud of depression swarmed over her, stinging like a nest of angry wasps.

Charity's voice broke into her troubled thoughts. "One more thing," she said, her tone softer yet filled with that familiar sarcastic edge. "If I ever catch you trying to put flowers in your hair again, I *will* tell Ma and Pa about your rebellion that night."

Faith's lips tightened in anger again, then softened when she heard Virgil's voice nearby.

"If you do anything of the sort, Charity, *I* will tell Ma and Pa what *you* did that night!"

Reaching one hand out, Faith searched in the air for her brother. It was a natural reaction. Touching, to Faith, was like seeing. Her fingers were her eyes. She felt him grip them with one hand. At the same time she heard Charity suck in a sharp breath.

"How dare you, Virgil! That's none of your business!"

Virgil laughed. "Oh, yes, it is, Sis, and if you threaten Faith again, I'll tell them. I mean it."

"What are you doing eavesdropping, anyhow?" Charity's tone changed to a whine. "I could tell on you for that!"

"Go ahead," Virgil said with indifference. "Let's see who gets in the worst trouble. I'll just bet Pa takes you behind a bush and whips you good." He let the words sink in. "Besides, I was only doin' as I was told. Dinner's ready, and Ma told me to come get both of you. I couldn't help overhearing."

At once Faith sensed the emptiness in the space her sister had earlier occupied.

"It's okay, Faith. She's gone." He touched her chin gently with his free hand.

Releasing his hand, Faith placed both palms on the ground and raised herself until she stood. Impulsively, she swished at her skirts, chasing away bits of grass, leaves, and tiny clumps of dirt.

"What else happened that night?" Faith asked, her

curiosity suddenly aroused.

Virgil hesitated, shuffling his feet in the patchy grass.

Faith broke the silence. "It's fine. I understand. You don't have to say a word."

"Yes," her brother replied, "I do. It's for your own good. That way, if Charity ever threatens you again, you have something to fight back with."

"But the Lord talks against being vindictive," Faith countered.

"Uh huh," Virgil said, "you're right about that. But I don't think anyone should take abuse off another person. You're blind, yes—but you're nobody's rug to beat."

Faith said nothing.

Gripping her hand tightly in his own, Virgil said, "I'll tell you. But not here, not now." He gave her a tug. "Come on, or Ma will pop us with that ladle."

He led her to the pot of stew, filled the tin plate, and handed it to her. Faith fed herself, without spilling a drop.

two

Virgil heard his mother order Charity down river to wash the dishes. Then she sent him and Joshua to check on the oxen grazing nearby and to grain their two horses, Nan and Nat. Nan would foal in a couple of weeks. Virgil hoped the trip hadn't affected the unborn colt in some way, as it had their milk cow. So far, the mare grazed and moved about contentedly, warding off the ever present affections of her admirer, Nat. Next Virgil checked the horses' hooves to be sure pebbles hadn't worked their way in between the shoes and the soft underbelly.

Of course, Joshua did little to help, to Virgil's continual dismay. The boy located a sturdy pine branch and made shapes in the dirt. One strap on his overalls fell off a shoulder. The blue cotton shirt he wore had scrunched up over his stomach nearly chest high. He seemed not to notice.

Peering back at the stock, Virgil sauntered to where his twin crouched near the ground. Joshua squinted up at him, the setting sun glancing off his gray eyes, making them appear translucent.

"Let's fix your shirt, Josh. That can't be too comfortable." Virgil tugged it back in place until it covered his brother's bulging abdomen.

"Pway wis me," Joshua begged. His deep voice sounded so much like a man's, but his words emerged childlike and garbled.

Virgil released a suppressed breath. He always took time to give Joshua the attention he needed; yet tonight he had promised *Faith* his time. "I can't tonight, Josh. Okay? But tomorrow I'll play soldier with you if you like."

For a moment it appeared Joshua might cry, but he didn't. He sulked instead.

"Okay, Josh?" Virgil repeated, praying there wouldn't be a problem.

The man-child stood and thumped the stick against the side of his head repeatedly. "Pwomise—you pway soder wis me damorrow?"

"I promise." Virgil gave his brother a big hug but took off running when Joshua threatened to return the gesture. His brother was stronger than any man he'd ever encountered. Virgil couldn't afford cracked ribs when his father needed him so badly to get the home built before the rains came. Quickly he helped James and his father sort the notched logs and stack them near places where they would be raised into position tomorrow.

Back at camp, he found Faith on a blanket laid over her usual spot on the ground, near their parents' bedding. That way Mrs. Lambert could help Faith if she needed to get up during the night.

Faith's hands held a book for the blind, a six-dot system, opened and worn around the edges from constant handling.

"You still tryin' to learn that stuff?" Virgil dropped onto the blanket beside Faith. He glanced in the direction of the wagon and saw his mother and Charity sorting kitchenware, bedding, and tools. They had hauled much of the wagon's load, including an old rocking chair and a buffalo skin hide, into the unfinished log home and secured it in a corner. The only thing left in the wagon was Mrs. Lambert's Saratoga trunk harboring precious family treasures.

In the distance James and his father discussed the next day's workload while Joshua made more scrawly lines in the dirt near the dying embers of the fire. Virgil didn't need any ears around when he said what he must.

Virgil wished he didn't have to speak ill of their sister, but Charity left him no choice. He loved Charity, but he loved Faith too. And what Charity had done was wrong! This might give Faith some weight in her situation with her sister. However wrong it seemed, there had to be something right in it.

He wondered why his mother and father hadn't noticed how badly Charity treated Faith. When they were around, Charity acted like the perfect sibling, doting on Faith like a mother. Of course they wouldn't know, mused Virgil.

"I want to learn to read," Faith confided, breaking into Virgil's thoughts.

"You know Ma and Pa would send you to one of them fancy schools for the blind if they could. They just can't afford it."

Running her fingers over several raised dots, Faith said, "I know. I don't blame them either. It's just—I want to read the way you do, like when you read Scripture in the evenings." She paused. "I bet Ma and Pa wish they could read like you and James and Charity."

"Maybe so," her brother agreed. "You know, Faith, in a way you *can* read. You can recite Scripture word for word better than any of us. You have some memory." When Faith didn't respond, he added, "Someday I'll make enough money to send you to one of them schools for the blind. Would you like that?"

Faith reached for him.

Virgil pressed into her open arms, allowing the gentle hug. When he moved away, tears filled Faith's eyes.

"I love you, Virgil, so much. I don't know what I'd do without you!"

Virgil loved his sister with all his heart, but moments like these made him edgy, like he itched all over and couldn't scratch.

"Let's talk about what we spoke of earlier, okay?"

"Sure." Faith's face grew somber, her gaze peering straight ahead as though looking through a glass vase and not at the object itself.

"I'm gonna tell you this, but I don't want you tellin' Ma and Pa unless you absolutely have to."

"Okay."

Drawing a deep breath, Virgil prepared himself. What he had to say would be difficult. "That night at the barn dance, when Charity slapped you," he paused, "she did somethin' she oughtn't to have."

"I know she shouldn't have slapped me. That was wrong."

"No, something else too. She did something else that night, and I caught her."

Faith's eyes widened. "What, Virgil? What did she do?"

Virgil hesitated. "I'd been watchin' Ma and Pa dance. They were laughin'. I'd never seen them let loose like that—you know, havin' fun. Then I realized I hadn't watched you girls, like they asked me. They asked James to as well, but he was surrounded by a company of smitten females." He chuckled. "Always the way with James."

When Faith didn't smile, Virgil picked up the dialogue.

"Josh was bein' watched by Ma and Pa. When I couldn't find you inside, I went outside. Found you in the wagon. Remember?"

Faith clasped her hands in her lap. "After I told you what happened, you tried to get me to go back inside with you. I wouldn't. I wanted to stay in the wagon until the dance was over."

"Only 'cause you'd been cryin'," Virgil said. "Guess that's what made me so mad—that and the fact Charity had slapped you." He sucked in his breath. "I could see her hand print across your cheek with only the light comin' from the barn. I can still see it if I close my eyes.

"Anyhow," Virgil continued, "I went lookin' for Charity, hunted all over the area, even went back inside and searched. Nothin'. I was returning to check on you when I heard it."

"Heard what?"

"The giggling. I knew it was Charity. Then I heard the other voice—it was male. I followed the sounds to a wagon set off away from the others, and there I saw them."

Faith's hands flew up to touch her cheeks, as though the motion provided a measure of comfort. She dropped them just as quickly. "N–no," she muttered.

"Yes. Charity was in there, and so was the man. I think he was older than me. He was quite large, larger than our Josh."

"What did you do?"

"I jumped into that wagon and yanked him by the arm— took him by surprise! Charity screamed and grabbed for her boots, but she couldn't find 'em in the darkness. Didn't matter none to me. My sister was comin' outta that wagon, shoes or not."

"What did the man do?"

"Laughed mostly. Called me a 'squirrelly kid' who would be 'reckoned with one day.' I offered to 'reckon' with him right then and there, but he wouldn't."

"He could have hurt you," Faith said, a worried look crossing her face.

Virgil chuckled at the memory. "Maybe so, but sometimes good reason doesn't always win out. I'll take care of my family—do anythin' for 'em. But Charity had strung my patience as far as it would go." He shifted on the bedding to a more comfortable position. "That fellow finally located her boots and tossed them out of the wagon at us. Charity put 'em on real quick. Then she started pleadin'." He grinned. "Oh, she was *so* afraid I was gonna tell Ma and Pa. She cried and threw all kinds of fits."

"Why didn't you tell them?"

"I didn't want to see her get in trouble, an' I knew she'd be in for it, if the folks got wind of what she'd done. Watching her shiver and cry for nearly an hour and all that pleading—well, I figured she'd learned her lesson. Told her if she ever did it again, things would be different, *much* different." He hesitated. "Before I let her go, I warned her never to slap you again."

"I–I never thought Charity would do such a thing."

Laughing, Virgil playfully stroked Faith under her chin. "Charity is different from you, Faith. You know that!"

"I know she's different from me, Virg. But she still should follow the teachings of the Bible. We were both brought up the same way."

"Doesn't mean it's gonna sink in with Charity, like with you. Charity may *hear* the words, but does she *listen?* Does she even believe?"

"Sometimes I wonder," she admitted. "I'm worried about her, Virg."

Virgil squeezed Faith's shoulder gently then retreated from the touch. "Charity has this wildness in her, like she's bursting at the seams with it. She wants to explore everything, even if it's wrong. Charity is stubborn too, and she's so fidgety. The girl can't sit still for a minute."

"She gets bored easily," Faith agreed. "But that's our Charity." She hesitated. "Do you think she'll come around?"

"Don't know. But if she doesn't grow out of this childishness, I'm afraid she'll have to pay the fiddler someday."

"We all pay the consequences for our sins," Faith said, knowing those consequences hurt at times. "God disciplines us because He loves us, just like Pa does."

Virgil didn't respond. A new voice broke into the conversation.

"We'd best git our readin' in before the sun's gone completely," said their mother. "Virgil, stir the fire a bit and add more kindlin'. I'm gonna round up the others." With that she was gone.

Before long, every family member was on his or her bedding. Virgil retrieved the Bible from the Saratoga trunk in the wagon, assuming he was going to read. James had read the previous week, and Charity the week before that, in a cycle that continuously passed between the three of them.

"Charity, you start reading for us tonight," Mrs. Lambert announced.

A groan rose in Charity's throat. "But it's not my turn!"

Grady Lambert replied sternly. "Do as yer told. Now!"

Virgil handed the volume to his sister and settled back on his bedroll.

Aware that her parents wouldn't know where she was reading from, unless they were told, Charity decided to choose her own spot. She didn't wish to continue James's readings from Isaiah. Scripture left a burning sensation in her stomach, though it would usually disappear by morning. She started in Colossians.

"Mortify therefore your members which are upon the earth; fornication, uncleanness, inordinate affection, evil concupiscence, and covetousness, which is idolatry."

Charity's heart leaped in her chest at the words. She flipped the pages quickly to another place.

"Flee also youthful lusts: but follow righteousness, faith, charity, peace, with them who call on the Lord out of a pure heart." Again the girl thumbed past the book of Timothy.

If she could have her say, she'd tell her folks she didn't believe God even existed. When you die, you're dead—that's it, she had concluded. What good does knowledge do you then? A person might as well live life to the fullest and experience everything, even if it's bad. *We only live once,* she would tell herself.

"I can't see that well." The girl closed the Bible. "It's too dark." She handed the book back to Virgil.

"Might be," said her father. "Sun's nearly gone. Might be we waited too long."

"I'll read some," Virgil said. "I can still see fine."

Mr. Lambert nodded approval, and Virgil turned to Isaiah 41, where James had left off the night before. He began reading. Before long, the waning light made him squint to see the words, so he decided to finish at Isaiah 42, verse 16.

"And I will bring the blind by a way that they knew not; I will lead them in paths that they have not known: I will make darkness light before them, and crooked things straight. These things will I do unto them, and not forsake them."

Closing the volume, Virgil could have no idea the impact this final verse would have on Faith. She had memorized it as he read and now clung to it, feeling it was meant for her, yet not understanding how it might apply to her personally.

"How 'bout some music?" James suggested, rubbing open blisters that had formed on the palms of his hands.

"Yeah, Virgil!" exclaimed Charity. "Play your guitar and sing us some songs."

The elder Lamberts nodded in Virgil's direction. Josh was already asleep when Virgil strode to the wagon's belly and retrieved his guitar. He returned the Bible to the safety of the Saratoga trunk.

The rest of the evening was spent strumming and singing around a dying fire. They sang "Oh! Susanna" and "My Old Kentucky Home" among other favorites. One by one, each member of the Lambert family tucked themselves into their blankets and fell asleep. Virgil was the last, or so he thought, as he laid the guitar down and settled himself back to welcome the night.

He knew tomorrow would be harder than the previous

week. Pa wanted to raise the rest of the home in one day, but Virgil didn't think it could be done. He figured they'd need at least three hard days or more. The roof alone would take that long. Finally he drifted off to sleep and dreamed of Nebraska.

Faith didn't sleep. She lay there, eyes wide open, unseeing. She could smell something, though. The scent was familiar, yet oddly strange.

To her it smelled similar to the Indian with a penchant for sour mash. Yet there was no scent of whiskey added to this odor. Mingled scents of leather, sweating horses, everything wild, and everything human reached her. Was she imagining it? Faith wondered. Should she tell someone? She heard Charity groan in her sleep; then she rolled over.

From the journals of Evan Dobson:

September 17, 1869, dusk.

I've shared company with a surveying team for the past two months. Had a good dinner of fried ham, green corn, and fresh milk. I find this occupation extremely interesting. Am considering giving up logging and rousting my own surveying unit. Mr. Carlton is very informative and answers my questions without hesitation. He told me to get in the practice of keeping a journal. So here I am, writing my first entry. Watched him change the base of the astronomical transit to east and west and set the Azimuth screw 116 points. Mr. Carlton takes plenty of sketches of the land to add to his notes. He also collects various insects and watches for fossils.

*Again I find my mind drifting back to Faith. It is for
her I make this trip. When I tire of the relentless travel,
I think of her and feel invigorated. Thought I might
catch up to the Lamberts but have run into problems
along the way. The old bay I started out on died. To be
expected as the poor thing was twenty years old and
the trip was just too much to endure. Some of Mr.
Carlton's company helped me rope a wild mustang.
Spent quite a bit of time breaking this beast. Now I've
got me a good horse, young and full of spirit. If it gets
me to where Faith is, I will be forever grateful.*

*My only concern is that Faith might not feel the
same way toward me as I feel toward her. Been on
my knees, praying nightly, for things to work out.
Maybe a little courting on my part could mean the
difference. Will close now. Intend to read* Harper's
Monthly *and then try to get some good shut-eye.*

The sound of her mother's voice and her brother Josh's
cries brought Faith out of her dreams, dreams of Evan,
where only sound, touch, and smell existed. They had
danced again, and he had kissed her. Faith was disap-
pointed at the interruption.

"Ah wet," Josh complained. "Ah wet!"

"Let's go down to the river, and Mama will fix it," his
mother told him.

Still Josh whined as he followed his mother down to
the river. Grady Lambert was already up, notching more
logs. Virgil and James stirred. Still Charity slept. Faith
knew by past experience that she had tugged the tattered
quilt up over her head.

Lifting her arms from beneath the blanket, Faith felt the day. Moisture permeated the air this morning. She reached out and patted the grass. Dew clung heavily to her fingers. Next she drew her fingers up to her nose and sniffed. A familiar, pungent scent greeted her yet was still too subtle, she knew, for the others to smell. She placed her arms flat against the blanket. There was little warmth, more on the cool side, which meant something blocked the sun. "It's gonna rain."

"Don't say that," Charity complained from beneath the covers. "I hate the rain!" She popped her head out and squinted brown eyes toward the sky. "Clouds," she groaned. "How will we all fit in that stupid wagon?"

"How does she do that?" James asked no one in particular concerning Faith's unusual gift. "Charity, get the fire going," he ordered. "We'll need some breakfast before long."

"And help Faith," Virgil added, "since Ma's with Josh."

At this, Charity bolted upright, her dark hair a stringy mess around her head. "Let Faith help herself," she retorted. "I have to do everything around here, especially for her!"

Faith swallowed hard. She hated her dependency—having to rely on everyone for *every* need. "It's all right. I'll manage on my own."

Charity laughed, and Faith cringed inside.

"Sure you will, little mouse!" she said scornfully.

Heat rose in Faith's cheeks. "I will! I don't need your help, not any of you!"

Kneeling beside his trembling sister, Virgil settled a restraining arm around Faith's shoulders. "Now look what you've done!" he shot at Charity.

The girl no longer laughed. "I'm tired of doing everything for her—chores, bathing, dressing! I'm tired of it. I'm tired of her!"

"Stop it!" warned Virgil.

Charity wasn't finished. "It's always Faith this or Faith that! Faith gets all the attention, while I do all the work!"

Peering eye to eye with Charity, Virgil said, "That's not true, so stop complaining!"

"What's going on here?" Mr. Lambert's gravelly voice silenced them. "Faith? What's the problem?"

"Nothing, Pa." Faith tried to control her trembling.

Mr. Lambert turned to Virgil. "Son?"

"You'll have to ask Charity." He stared at his sister.

As her father approached, Charity peered at the ground.

"Daughter?"

For a moment Charity said nothing.

"Now I heard voices raised, and somebody better git to tellin' me what the ruckus was about!"

James, off in the distance, notched logs with the axe, obviously wanting no part of the squabble.

Suddenly Faith spoke up. "It's my fault, Pa. I didn't want anybody helping me." It was the truth. She wished that nobody had to do a thing for her, *especially* Charity.

"Don't see as ya have much choice, Girl," her father said sternly. "Now ya best see to watchin' that temper of yers, or I'll take a switch to ya. Understand?"

"Yes, Sir."

Still Charity said nothing.

"Virgil, we've much work ahead. Let's git movin'." The older man lumbered back to his building project.

When he was out of earshot, Virgil spoke. "You let her take the blame, Charity." He kept his voice low. "That ain't right!"

Shrugging, Charity muttered, "I didn't ask her to."

"You just better watch yourself 'cause Faith knows what you did that night."

"What?"

"That's right! She knows everything. Told her last night." He peered down on Faith. "I'd be extra nice if I was you." With that he stomped away to begin the task of stacking logs.

Heavy silence enveloped Faith and Charity. For a long moment neither moved. After several minutes Faith heard the rustling of skirts. Charity knelt beside her.

"We'd better get your hair combed," she stated simply.

"Just hand me the brush. I'll do it myself."

Charity's voice grew sickly sweet. "I'll do it up real pretty for you," she insisted.

"I said I can do it *myself.*" Faith no longer trusted her sister. Something had changed between them.

It had started a few years back, when Faith had grown into a woman. At first, she sensed the subtle signs of their relationship becoming uncomfortable, like a tiny cut. When the cut wasn't treated, a wound developed and festered. Not long after that, infection set in, and soon it would be gangrene and out of control.

Faith believed her relationship with Charity was at the infection stage now, especially with what she'd learned last night from Virgil. No, she wouldn't tell on Charity. It would be wrong to use that against her, although she believed their mother and father had a right to know. It was Virgil's responsibility. Right?

Confusion filled her. She loved Charity! Why must it be like this? She felt the brush as it landed hard against her palm.

"Thank you." She placed the bristles against the crown of her head and worked them downward to where the ends of her blond hair reached her waist. With her free hand Faith felt the stroked locks, making certain, as best she could, that they were smooth and free of tangles.

Suddenly, Faith felt the pressure of a cool, wet towel as Charity tried to wash her face.

"Let me do it," Faith said, reaching out for the wet towel.

Her anger had dissipated, but she knew in her heart that Charity was only concerned for herself. Faith knew the awful secret, Charity's secret. Her younger sister would do whatever she must to keep that information between the three of them.

"I'm only trying to get you ready for the day." Charity rose and then squatted beside her sister again. "Why are you being so stubborn? Let me help you."

Faith ran the cloth across her eyes, down her face and neck. She had relied on others for too long. Blind she was, but not helpless. Surely she could do some things, learn some skills. She just had to find them.

She heard Charity brush her own hair and straighten her mussed skirts with a *whoosh* before meandering down to the river to fill the coffeepot with water. The morning routine was ritual.

❧

The noon sun broke through the clouds, but only momentarily. Josh napped in the wagon. James, Virgil, and their father raised the walls on the home one log at a time while Charity and her mother applied chinking between the gaps once they were up. Portions of the roof were already in place.

Faith tried to wash the breakfast and lunch dishes for the first time. Her mother grew worried that Faith might lose her footing and fall into the river, so she made her stop. Not a word had passed between Charity and Faith the entire day thus far.

As Faith sat on her bedding beside the wagon, she ran her fingers over the raised dots in the worn braille book and daydreamed of other worlds.

Occasionally Joshua cried out in his sleep, and Faith would soothe him with her voice. He grew quiet again each time as though nothing had happened.

He had experienced nightmares since their arrival. Faith didn't understand the reason but thought perhaps the move had affected him somehow. She was worried about her brother. Something else troubled her though. The scent was back.

Having disappeared with the morning, she had given it no more thought. Now it had returned—that wild, pungent odor. What concerned her most was that it was human—very human.

Faith couldn't define human scent. She had smelled one part of it on every person with whom she'd come in contact. It managed to eke out from under a man's sweaty body odor or a woman's strong-smelling cologne. The scent knew no race. Nor was the smell entirely bad or entirely good—rather a cross between the two. Yet it was always present.

A man and a horse could both smell of sweat and leather. Faith could tell the difference between them easily. What she smelled now belonged to man *and* beast and contained a wild odor not generally possessed by the white man. Faith could smell them both distinctly. At times the breeze brought her a strong whiff. She must tell someone!

"Virgil!" she called out, forgetting about Josh sleeping in the wagon.

He cried as he wriggled from his sleepy stupor. Faith rose quickly and fumbled her way into the wagon to console him.

"Come to me, Josh," Charity demanded from somewhere behind them.

Josh only clung more tightly to Faith's neck, nearly pulling her down with his weight. "Pwivy," was all he said.

"Go with Sister," Faith said gently, wishing she could tend to his needs herself.

"No!" Josh insisted. "Me go wis Faith!"

Faith heard a breath of exasperation leave her sister's lips.

"Joshua, quit being a baby! I'll take you to the privy!"

"Don't yell at him," Faith said, stiffening. "You'll only upset him more."

"Ma told me to get him when he woke."

Faith tugged at Joshua's fingers to pry them off her neck. He was strong, and Charity had to help. When they succeeded, Charity gripped him by the hand and yanked. Josh yanked back.

"Go on, Josh," Faith told her brother. "It's okay."

Joshua began to cry. "Me stay wis you."

"Don't you have to go to the privy?" Concern drove an uncomfortable wedge into Faith's heart. Why was he so afraid of Charity?

"You take me," he begged Faith.

Faith wanted to help her brother, do things for him, be useful to him. A sense of hopelessness filled her.

"I can't," muttered Faith, her voice shaking. "Be a good boy now and go with Sister. She'll take you real quick, and then she'll bring you back. Okay?"

"Pwomise?" Joshua's favorite word trembled in the air.

Faith raised her chin, tilting her face in the direction where she knew her sister stood. "Will you bring him right back?"

"Yes," Charity said abruptly.

They disappeared into the thicket.

Again Faith called out Virgil's name as she climbed from the belly of the wagon and resumed her place on the quilt. A few moments elapsed before she heard his approach.

"You okay?" Concern tinged his words.

"Can you sit a moment?" Faith asked, forgetting to answer his question.

"Can't. Pa's in a big hurry to get the rest of the roof up. Says the rains are comin'."

"Just for a minute. Please?"

Virgil didn't sit but knelt close to his sister. "What is it? Charity causin' you problems?"

"No, but I've got to tell somebody what I smell. I smelled it last night. That's why I couldn't get to sleep. The scent is stronger this time."

For a moment Virgil was quiet. He ran a hand through his sand-colored hair and drew a deep breath. "What do you smell?"

"Indians."

Silence encased them so tightly that goose bumps freckled Faith's skin.

"You sure?"

Virgil asked the question so suddenly that Faith jumped. She drew in a deep breath of her own.

"Yes, and you must tell Pa."

Again silence encroached, as though Virgil was in deep thought.

"Okay." He rose to go.

Faith felt a measure of relief at the response. She didn't know Virgil trusted her abilities beyond doubt—even if the others thought her assertions were nothing more than mere coincidence.

Within seconds Joshua returned from the bushes. He crawled onto the quilt beside Faith and rested his head on her lap. His large body took up most of the blanket as he lay on his side. Joshua always woke slowly. Faith could hear him sucking on his thumb.

Before long, Mrs. Lambert approached. "Take this licorice and go play," she told the boy.

She heard his heavy, stumbling walk as he bounded off toward the others where they worked. Faith couldn't help but smile. Josh always seemed to need attention. Then she remembered her mother. Hearing skirts rustle, she knew she had sat down opposite her on the blanket.

"Ma?" asked Faith, though she knew full well it was she, with no words being spoken.

"It's me, Child. What's this yer tellin' Virgil 'bout Injuns?"

"Indians, Ma," Faith said, not understanding why a slang term could so easily be attached to human beings.

"What about 'em?"

"I can smell them."

For a moment Mrs. Lambert said nothing. Faith shifted on the blanket, uncomfortable with the silence and wanting to see her mother's face, her eyes, wanting it with all of her heart.

"Ma?" Faith wished she'd say something, anything.

"You sure 'bout this?"

"Yes."

"Well, I'll go tell Pa—see what he thinks." She rose to leave.

Faith strained to hear their murmurings. She knew her father didn't trust her abilities.

"Ain't no Injuns 'round these parts," she heard him say, with irritation in his voice. "Faith has a wild imagination. She's just nervous 'bout being in new surroundings. What happened before is over and done."

"What if she's right?" Mrs. Lambert countered.

"She's wrong!" her husband's voice was gruff. "If ya want a house up, you'd best let me see to my work."

The conversation ended. Mrs. Lambert didn't return.

Resting on her back, Faith allowed the scant afternoon sun to console her with its warmth. She wanted to nap but couldn't. The scent of Indians wouldn't leave her. Fear rose up strong inside of her. No amount of sun could chase away the chill in her blood. And then the rain came.

three

Another day passed without incident. The scent disappeared. Faith relaxed. But she vowed to remain alert, and Virgil promised to keep watch.

The evening meal had been meager. Mrs. Lambert had no flour, coffee, or beans, and few other essentials. Mr. Lambert had offered to ride to Del Norte when the sun rose. He wanted Virgil to accompany him. The thought of Virgil being gone unnerved Faith. She told him about her concerns while the rest of the family slept.

"Virgil," she whispered. Usually they were the last to fall asleep.

"What?" Virgil pulled his blanket off and crawled to where Faith lay, close to their parents.

When she was certain nobody stirred, Faith spoke. "I don't want you to go tomorrow. Can't James go instead? Pa usually takes him, not you."

"He just wants James to stay behind this time so he can keep raisin' logs. Pa's in a hurry to get the home built so we can move in before more rain comes." He paused. "Go to sleep, Faith. Everything'll be fine."

Virgil moved quietly back to his pallet to settle down for the night.

Faith sniffed the air. Nothing. Maybe the Indians were only curious and had ridden on. The rhythmic sounds of her father's snores calmed her like a lullaby. She never felt her eyes close, when they did at last, three hours before dawn broke.

When she awoke, the sun had scarcely shouldered its way above the rim of the horizon, casting about a warmth that mingled with the cool morning mist. She knew her father and brother had gone. She listened as the remaining members of the Lambert family completed the sunrise ritual and set to their tasks, which included baby-sitting Joshua.

Faith ached inside when she heard the swift strokes of the axe as James notched logs and the rattling of tin cups, plates, and utensils as they were being washed by the river. She longed to be part of the activity, part of life, instead of being sheltered in her cocoon of darkness. She was allowed to complete only the most elementary tasks, no more. These thoughts fled her mind when the scent returned.

Rising to her knees, Faith placed both palms on the ground in front of her. She drew in the smell, her nostrils flaring with the effort. Too close! Had to be right upon them!

"James!" she screamed. The chopping halted, but no response came.

"Ma!" Faith yelled when James didn't answer.

Then she heard her mother shriek and Joshua yell "Mama!" again and again.

"What's happening?" she cried out in James's direction.

She heard no words, just a sound as though he'd been punched, perhaps in the stomach, knocking the wind out of him. Finally he spoke. His words made Faith dizzy with fear.

"Run, Ma! Joshua, run!"

"James!" Faith grew frantic. The sound of horses and footsteps surrounded her. She gripped the quilt she'd been sitting on, holding it close to her chest, her forehead creased with terror.

"Get to Faith, Ma!" James shouted again. "Get—inside the—wagon!" His words seemed strained, as if he'd had to force them out.

Suddenly Faith heard a soft whir, as though a concentrated breeze gusted nearby. It was followed by a thump that emanated from part of the wagon. Struggling to her feet, Faith fought to climb into the shelter of the wagon. She caught her foot in the hem of her skirts and fell back out.

"Help me!" she pleaded.

She heard heavy breathing beside her, then her mother's voice.

"Get into the wagon, Girl! You too, Charity!"

Faith felt the sudden breeze as her sister ran past her. Mrs. Lambert pushed her older daughter onto the seat and then inside where she tumbled on top of her sister. She righted herself at once. A gun blast sounded nearby, from James's direction.

"Ma? You coming?" Faith reached into the air, desperately searching for her mother. She heard Charity sobbing and realized her sister must have dropped behind the Saratoga trunk, probably curled into a ball.

When no answer came from her mother, Faith called out a second time. "Ma? You coming?"

"Have to get J—Josh," her mother muttered.

Josh! Where was he?

"Ma!" Too late. Faith felt the emptiness. She reached toward the sound of Charity's cries and grabbed her sister by the arm. "Joshua! Where's Josh?" They had to help him. He was defenseless!

Faith searched for the crate holding the family's kitchen supplies, seeking a sharp knife, a weapon. When she couldn't locate the crate, her mouth went dry. She remembered it had been unloaded and was now under the wagon to provide more room inside in case they needed shelter from the rain.

She needed Charity's help but knew the girl was too frightened to move. Faith must act alone.

In her darkness she located the wagon opening and maneuvered through it and out onto the buckboard seat. James must have seen her.

"Faith, get in the wagon!" he cried.

For a second Faith ducked back inside, her heart beating furiously. *I must do something!* "Lord," she said aloud. Charity's sobs nearly drowned out her words. "Keep watch over me." Then she was back outside. Before she knew it, she was on the ground.

"No!" James insisted again. "Get back—"

His words were cut short by Mrs. Lambert's scream. Sounds permeated the air around her. Grit found its way into Faith's eyes and mouth as dirt flew.

The tears forming in Faith's eyes helped wash some of the grit away. She heard footsteps approach and probed the air with her fingers. "Who's there?"

"It's me," her mother said, her voice choked and low, lower than Faith had ever heard it.

To her relief she heard Joshua crying as her mother

rushed him to the safety of the wagon. At her mother's next words, relief vanished.

"Josh! No! Come back to Mama!"

"What? Where did he go?" Faith's heart beat faster.

She heard something split the air and again heard a thud, softer this time than when it hit the wagon.

"Oweee!"

Joshua. Faith's knees nearly buckled under her. He must have been hit with an arrow. And James? What of James?

She could hear footsteps running away. Her mother must have chased after Joshua. Faith stumbled after her, falling over yucca, cactus, and thick clumps of bunch grass. Suddenly an arrow cut the breeze close to her head. Another thud, the most horrid sound Faith had ever heard, and a moan. She found her mother and Joshua on the ground, very still. Her breath caught in her throat.

Another arrow sliced the air to her left with a soft hum. She would never forget the sound, like a sighted person might remember a horrifying image. Faith shoved herself back, afraid it had been aimed at her. Then she remembered the prayer. "Keep watch over me." Her beating heart threatened to pound the words from her memory. Yet they were all she could hold onto.

Hoots and howls filled the air, mingling with it as though part of the wind itself. A warrior laughed from the direction of the cabin, where James had been working. He must be wounded, she thought, or—

Faith stood and followed the sounds of Charity's terrified screams. She struggled into the wagon.

"What good does the wagon do?" cried Charity. "Nothing will stop these savages from killing us all!" Her voice choked with fear. "Can we escape?" she asked.

"No," said Faith, "it isn't safe outside." Then Faith did the only thing she knew to do; she bowed her head and prayed softly.

"What good are your prayers? We're all going to die!"

The yelping rose in crescendo.

Charity crawled back to the trunk. "It's over," she said between deep sobs. "It's over."

The muffled thud and smell of smoke forced Faith to regain her focus.

"Smoke," Charity said.

A fiery arrow pierced one side of the trunk that held all their mother's earthly possessions. Faith could feel it and smell it.

"The wagon's on fire! We have to get out!"

Faith fumbled her way to the trunk and lifted the lid. She gripped the family Bible and threw it, not knowing if it escaped destruction or not.

Flames consumed the osnaburg, traveling fast over their heads. The smell was overpowering, but the heat was worse. Charity screamed as fiery tongues licked close by.

"Come on! We have to get out!" Faith urged her sister again.

"They'll kill us!"

"We don't have a choice."

The decision was made for them when the heat grew unbearable.

Faith heard her sister groan when she landed hard on the ground outside. Faith followed at once.

"The osnaburg is completely gone, destroyed!" Charity

cried. "And the edges of the wheel rims are glowing red hot—like they're in a furnace."

Unable to understand, Faith edged closer to Charity, feeling the horrid heat coming off the burning wagon. Laughter permeated the air around them.

Suddenly Charity cried. "My hair's on fire!"

Faith moved at once and found the source. She put out the blaze with her bare hands, leaving behind rising blisters.

Breathing heavily and rubbing her head, Charity looked around. The Indians had them surrounded. Some were mounted on war ponies. The horses had been painted red. Others left their mounts and stood gawking at the two white women.

The warriors wore hide flaps hanging from a belt in front and leggings drawn tight above the knee, yet loose below. Painted red and black, they were decorated for war. Many gripped lances that held bloody human scalps. They waited.

Soon the fire burned down. All that was left of the family wagon was smoldering ashes and smoking iron remnants. The crate beneath the wagon was destroyed along with its contents. A heavy silence filled the air. Charity and Faith stood side by side covered in soot and dirt, their dresses singed and ripped.

Charity struggled to keep from looking at her mother's lifeless body and those of her brothers. She kept her eyes downcast and avoided the smirking glances of the warriors encircling them. For the first time in her life she wished she were blind, like Faith. The carnage that lay before them was more than she could bear.

One man approached Charity and gripped a section of her long, dark hair. One small portion had been singed nearly to the scalp. He spoke a strange dialect. His words brought laughter from the others.

"Please," Charity begged.

"What?" asked Faith, moving toward her sister.

"Don't," Charity warned her. "Stay where you are."

Faith touched Charity around the head, her face tilted upward, as though studying the sky. As Faith reached for the source of discomfort, the man dropped Charity's hair and moved away.

This time nobody laughed.

"How did you do that?" Charity asked.

"I didn't do anything. What happened?"

Her breath came in spasms. "It's like he's afraid of you."

Stepping closer to Faith, she said, "That's why you weren't struck by arrows when you went after Josh. They think you're touched, Faith. They think you're crazy!"

"I prayed for protection."

"No," Charity argued, "they think you're crazy."

"You heard me pray."

"Yes," Charity responded, peering around at the on-lookers, "I heard you pray." She couldn't deny that.

Faith spoke again softly. "God just chose this as His way to protect me."

With no rebuttal Charity fell silent. A second later she asked, "What do we do now?"

"Stay by me. If they think I'm crazy, they won't touch me—or you. Remember the story of the man who escaped a band of Apaches because he acted crazy?"

"Yeah, Pa told it to us. The Indians think it will affect

them, bring evil spirits on them and their families if
they touch a crazy person." She spoke the words fast,
as though they might be her last.

"Right. So stay by me."

Horses snorted, and the warriors riding them moved in
for a better view. In an effort to keep her distance, Charity
stepped backward too far.

One warrior rode close and, with one motion, slung
Charity on top of his horse in front of him. She struggled
to free herself but couldn't.

Faith thought her sister to be nearby, though she could
no longer feel her presence. Silence engulfed them. Then
she heard the moaning and knew it came from Charity.
She had been hurt, wounded, maybe knocked out.

"Charity!" she shouted. "Are you all right?"

There was no answer. Faith instinctively secured both
palms to each cheek, not realizing she did so.

Again she called out for Charity. Another warrior approached Faith. When she felt the heat pressed close to
her eyes she realized someone had passed a hand in front
of her face.

The hand passed by several more times. Sounds, words,
shot through the air. Faith knew her secret had been discovered. Like Charity she was whisked atop a pony, in
front of her captor. The wind threatened to suffocate her as
they rode into it. She fought with all her might to hold on.

four

At times Faith believed she might tumble off the horse. She struggled against her captor—not wanting his nearness. Always she felt his bruising grip to secure her back in front of him on the buffalo-hide saddle. In her dark world she felt the continual scrapings against her legs, arms, even her face. She could also feel his long hair when the wind whipped it a certain way and it landed with a sting against either of her cheeks. Tears rose in sightless eyes, though none fell. The wind dried the salty drops as soon as they sought freedom.

She ached everywhere. The journey seemed endless. The only thing familiar was the beating of her heart, pumping with fear, telling her she was alive, over and over. What torment lay ahead? And what would become of Charity?

Thinking of her sister caused her to struggle subconsciously. Again she was positioned upright with a stern warning.

"Hekotooestse!"

It was a word she did not recognize, but one she had heard often in the last several hours. Had it only been hours? She thought an eternity had elapsed since the nightmare began.

A branch slapped against the soft nestle of her neck, its hardness leaving behind a bloody scrape. Sweat surfaced

49

from fear and the hot sun pounding down on them as they rode at breakneck speed. Salt from the sweat stung the wounds, telling Faith they covered nearly every part of her body, even the midsection where her garment should have offered protection but didn't. The blisters on her hands from putting out the fire in Charity's hair burned with even greater intensity.

Branches from trees and prickly bushes sliced into the gingham, carving her flesh with the deftness of a sharp dagger.

Faith tried to shut herself off from the event, like her sight cut off bad things she appreciated not seeing. All her other senses worked well, more than well; they burned each sensation into her like a branding iron placed against living skin.

The hot wind offered no solace, only difficulty in breathing. The scent of the Indian was strong, pungent almost, and wild. They remained quiet but rode fast. The ground beneath crackled, probably from dead leaves and pine needles. Lashing noises radiated from higher. Whips were used to goad the horses onward. These sounds seemed to swell up around Faith, almost overtake her, even though they were natural in every sense.

She had heard them before, while her family traveled over the past year. Now they seemed different, as if they had a life of their own and a tireless fury.

Faith felt utterly lost. She didn't know which way they were traveling, although she had learned to judge direction and time by the sun, the way it warmed the body. Faith was never given the opportunity to test the sun's warmth. She whispered her sister's name.

"Charity." The word lingered on her lips like a voiceless cry. Faith's chin quivered as another branch sliced near her right temple. Tears fell but were obliterated by the rushing air, as though the wind warned her *not* to cry.

Her voice knotted in her throat, leaving behind a giant lump through which she could not scream. Trying several times to believe it was a bad dream, Faith was hurried back into the present by the painful gouging of a branch or the scraping of a stiff twig only to feel agony stronger than the last. She tried to think of survival in a reality beyond the human experience.

Then greater fear coursed through her veins. What would happen when they reached their destination? Surely they wouldn't travel like this indefinitely. No, they would reach a point for rest. The horses could run only so long. Then what? Torture? She'd heard the stories, like everyone else, generally around a campfire when all was safe and guarded, when life was good. Faith chose not to believe the stories, that anyone, any human being, could ever commit such atrocities against another. Was she wrong?

"God help us," she whispered into the wind as it choked her air away.

Before long, Faith felt the warmth of the sun abating and the coolness of night beginning. She knew it was early evening. Still the band pushed onward.

She wanted off the horse so badly that she almost decided to roll off at a certain moment, hoping the fall might knock her unconscious, maybe even bring the release of death. But no—dying by her own suggestive will had never occurred to her before, nor would it. Faith understood now, though, what brought people to that point.

She was in the abyss, walking in that dark pit this very minute.

More than ever she had to believe God was with her. She *had* to believe.

Soon the night chill moved across her skin, making bumps rise. When would this relentless riding cease? Faith would accept her fate, if only they would stop. She ached from riding so long; she was weak and tired and didn't know how much more she could endure.

Earlier, the band had ridden with a vengeance. Now they moved slowly, quietly, as though hiding from something, fearful of it. They never traveled by the river, though they had stopped once to fill buffalo bladders with fresh water. Then they shoved back into the thickness of trees, brush, and thicket.

Faith grew sleepy. Exhaustion overcame her, and she struggled in vain to stay awake.

She woke with a cry. She had plummeted from the horse into a patch of thorny branches. One limb had gouged her upper arm. At once an arm encircled her waist, and a hand pressed over her mouth and nose with such force that she fought to breathe.

As if the warrior realized, the hand dropped enough to release her nose, allowing her to inhale. Faith stopped struggling and waited, anticipating the worst. Harsh words were spoken, this time near her ear in a whisper, yet with the same fury as before. She knew not to scream again.

Words flew in sharp whispers around her, like horse whips that might strike her at any moment. Her heart beat faster, and she struggled to contain her emotions. Silence followed. Then Faith heard several feet hit the ground as the warriors apparently dismounted. Some of the horses galloped away, while others remained.

Charity cried softly somewhere to Faith's right. Anger burned inside her, chasing the chill away. All right, thought Faith, let them do with her as they wished. She would die without a struggle—something she sensed they wouldn't like. If her time had come, she would go bravely to be with her Lord.

≥•

The oxen trailed the two horses Mr. Lambert and Virgil were riding. Burlap sacks and hide *parfleches* holding nonperishable goods were tied securely to their backs. Some dangled from ropes on each side of the oxen.

They reached Del Norte by midafternoon, hurriedly made their purchases and set off again. Virgil knew his father's thoughts were on the log house, though Virgil would have enjoyed talking with some of the occupants of the little town, especially the pretty women.

The Rocky Mountains loomed, though they could no longer be viewed. Darkness, lit only by the moon, shrouded the men as they journeyed. Virgil knew it would be late when they arrived home. He relaxed in the saddle and breathed deeply of the sweet scent of yarrow that permeated the cool mountain air.

Occasionally, they let the horses graze on clumps of bunch grass and the pastel pink and white blooms of a scattering of bindweed.

Mr. Lambert removed his tattered black Stetson and slapped it against his thigh several times along the way. Silver tufts of hair edged his temples. The rest was coal black like James's.

As they neared home, Grady Lambert's smile plowed deep crevices into the outer, fleshy corners of his eyes. Virgil knew his father had found his home in Colorado.

The way was slow, but all went as planned. Virgil was eager to see how the log home had progressed in their absence. His father wore an ever present smile on his face. To Virgil, he seemed content with the turn life was taking for him and his family. The future did, indeed, look good for the Lamberts, Virgil believed.

➢

The riders returned, almost as quickly as they had gone. A heated dialogue followed, and Faith grew concerned. *What's going on?*

Suddenly a warm, flailing mass hit her, a living body. Charity grabbed hold of her sister tightly. Faith managed to stay upright. They stood side by side, trembling.

"What are they doing?" Faith whispered.

"I don't know." Charity paused. "They're all getting back on their horses."

Faith heard the sounds of hooves digging into dirt. Branches snapped; leaves crackled. Then all grew silent. The air reeked with the smells of dirt mingled with leather and horse sweat.

Both girls inhaled sharply. They stood alone in the middle of nowhere.

"They're gone," Faith stated.

"Yes, I–I think so."

"Don't move just yet. Be still."

Neither moved for several minutes. Finally Faith spoke. "They left us."

"Why?" Charity's voice held hope, yet confusion.

Faith gripped her sister's hand. "We must get to shelter. Hide. Quickly!"

Charity found a bush thick with leaves. Faith allowed herself to be led, then helped into its covering where she settled into a sitting position. Charity ducked in, pressing close to her sister. Faith grabbed remnants of her torn dress and tucked them around her knees.

"Why did they leave us?" Charity asked again. "They didn't even kill us! I thought we were going to die!"

For a second Faith remained quiet, listening, smelling the wind. Yes, they were gone. For the first time her heart slowed its beat. She sucked in a deep, replenishing breath.

"I think the one who left may have seen something. Or—" She hesitated. "They think we're too much trouble, couldn't travel as fast with us."

Again she paused, smelling the air. "They might decide to come back though."

She didn't want to frighten her sister. Yet she had to be truthful if they were to survive. False hope might cause them to let down their guard prematurely.

"Faith! Oh, Faith!" Charity gasped. "What will we do?"

"I don't know. Did they hurt you?"

Charity released a jittery breath. "I hurt all over. And I'm scared, Faith. I'm *so* scared."

When Faith heard her sobs, she reached out, took her hand in her own and squeezed gently, ignoring the burning blisters.

"What if they come back?" Charity whispered. A tremulous edge sliced into her words.

"I don't know, Charity."

Silence covered them.

"Tell me what you see around us. The land. What does it look like?" Faith realized she would have to take control. But she would have to use her sister's eyes to do so. She would *not* just sit out here and die. Yet that overpowering sense of dependency bit at her heart like a venomous snake. She forced her thoughts back to the present.

Charity held her voice, as she studied their surroundings.

"It's dark," she said, "but I remember seeing lots of pine, spruce, pinyon, and juniper trees in the distance and sage brush. Yucca is all around and bindweed. Lots of purple lupine, yarrow, sweet clover, and other wild flowers."

Sniffing the air in response, Faith separated and discerned the various scents, already familiar with most. As she did this, she felt a measure of comfort wash over her, bringing with it a sense of control. She was using what she had available, instead of focusing on abilities she didn't have.

"The land is flatter here, I think," Charity continued, "and bare. Trees are mostly along the river banks and at the base of the mountains." She paused. "What will we do?"

"I'm not sure."

"Walk? Stay here? What?"

"I'm not sure," Faith repeated.

Charity began to cry.

Reaching for her sister's face, Faith stroked her cheek. Charity's tears clung to her fingers. "Shh," Faith told her. "I need you to be strong for me. Okay?"

Faith heard her sister draw in a weary breath.

"We have to be strong," Faith continued. "We can make it through this if we're strong."

"I–I can't. Mother. . .J–Josh. . .James. . . ."

"You must." Faith still gripped Charity's hand in her own. She squeezed against the pain.

"Pray with me," Faith asked quietly.

Charity didn't argue. This brought Faith a new measure of hope, hope for Charity, hope for her change of heart.

"Father in heaven, we ask for Your help. We ask that You be with us, that You keep us safe from harm. We ask that You lead us safely home. And—and please, Father, keep Virgil and Pa safe. Help them—" Her voice faltered. "Your will be done. In Jesus' name we pray, amen."

As her voice wavered on the last word, a verse came to mind, the same one in Isaiah that Virgil had read around the campfire. Faith recited the passage aloud, letting the words cloak her like a warm blanket on a chilly winter's night. She knew the Lord was with her, bringing comfort.

"And I will bring the blind by a way that they knew not; I will lead them in paths that they have not known: I will make darkness light before them, and crooked things straight. These things will I do unto them, and not forsake them."

It dawned on Faith that Charity's heart couldn't *see* any better than her own eyes. Her sister was blind in a different way. But she couldn't dwell on this revelation right now.

"Do you have any idea which way we were traveling?" she asked.

"No."

Quiet filled the space between them. They would spend the night here and nurse their wounds as best they could.

Faith recited the passage over and over to herself, refusing to allow fear to consume her again. But what if they came back?

❧

Evan Dobson watched the riders from a distance. When he was certain it was Grady and Virgil Lambert, he coaxed his white mustang into a slow trot.

His heart thumped with the steady rhythm of a sweet melody in his chest. Soon he would see the lovely Faith again, the woman for whom his heart sang, the woman of his dreams. As Evan rode, he remembered.

Growing up in Prague, Nebraska, Evan had always been fascinated by Faith. While many avoided her, he felt drawn to her. But that pesky little brother of hers, Virgil, wouldn't let anybody near his sister.

He understood that to a degree. The children at school teased Virgil's twin, Joshua, with a brutality Evan couldn't comprehend. Virgil probably felt Faith couldn't endure such taunts. He protected her with a vengeance.

Every day Evan entered the one room log school building with its hard wooden benches and cast iron stove to learn and to watch Faith. Yes, he could study her without fear of upsetting her. As she grew, she became more beautiful. Those ice blue eyes only enhanced her perfectly sculptured features. High cheekbones, straight, narrow nose. And lips, bottom slightly fuller than the top, curved into her creamy complexion, adding a quality to her beauty he had never seen in a young woman.

To him she seemed perfect, except for her lack of vision. And that didn't matter, then or now.

He had observed her sadness at this fact, at her blindness—especially when he and other students learned to read from the McGuffey's Readers. Faith couldn't read or write like the sighted students, but she had a memorizing ability that surpassed all.

She memorized every grammar rule they were taught and could spell out every word, though she didn't know the full meaning of her efforts. Regardless, she entered the spelling bees and won. He remembered the teacher, Mrs. Thornton, asking Faith if she understood what she had done. Faith explained that she did not know the letters by sight but by sound. She "read" by sound and memory. Astonishing!

Faith recited history dates with an accuracy that escaped the rest of them. But her greatest talent lay in telling stories to the class.

Story time was right after lunch. Having memorized every story the teacher read, Faith was asked repeatedly to perform the task. She recited each one almost word for word. That's when he saw her light up, as though the sun penetrated her being and shone its golden veil through her. That's when she seemed happiest.

And her voice. It's soft, smooth tone had a soothing effect on him and on the others. He could tell. When Faith told stories, the pupils grew quiet, as though mesmerized, either by the story or by her ability, or both.

No, Evan couldn't understand the irresistible draw he felt toward Faith nor could he deny it. They had grown up together, though he had scarcely uttered a word in her direction. She seemed untouchable, and he lacked the confidence needed to approach her at the time. He could only watch from a distance and fall more in love with the girl as each day came and went.

Then the dance came. Something twitched in his stomach when he remembered the barn dance. He nearly laughed out loud. Charity, though she tried, couldn't stop Faith's natural beauty from shining through. It was an internal thing one couldn't staunch. Faith was not Charity's rival, though Charity chose to believe so, for Faith didn't even know she was beautiful!

He found irony in the fact that Faith was the fairest woman there; she stood out and was incredibly pleasing to the eyes. Yet she couldn't even see through her *own* eyes.

When he took her small hands in his large ones, he marveled at the fact that he was actually touching her. *Finally* touching Faith! How long he had waited for that!

Time seemed to stand still as he danced with the woman he knew would be his wife someday. Evan knew this without a doubt. His only concern was that Faith didn't know it yet.

His joy was short lived when Charity chose to cause a scene. Her actions appalled him. The unfairness of it all was something he could never accept.

Ever since he could remember, Charity had been wrapped up in herself, in her own needs. She had never given Faith credit for anything. Only God knew what Faith had endured, then and now. He would soon change all that, with God's help.

He was rushed back to the present when he caught sight of something in the distance. The Rio Grande river flowed nearby. A house, he presumed, stood silhouetted against the horizon. But an uneasiness caused him to peer closely in that direction.

Placing a thumb in his belt loop, Evan allowed his hand to go limp. Quietly he goaded his horse onward. *I'm overreacting,* he told himself. *It's been a long jour-*

ney. It's late, and I'm tired. Yeah, I'm imagining things.

Something gnawed at his gut. Evan never doubted the voice that spoke to his heart. God had never steered him wrong. He pressed forward at a slow canter.

❧

Virgil felt his heart skip a beat as they approached. Something wasn't right. He glanced at his father and saw a look of foreboding on his face.

Smoke. No, not smoke, but the scent of something charred reached Virgil's nostrils. Where was everybody? The silence over the place was unsettling.

Just then Virgil saw it. James. And Joshua. Sleeping? A knot formed in his throat.

Mr. Lambert dismounted first. Virgil stepped down from the saddle, leaving Nat's reins to drag the ground. The oxen would contain the beast.

Again he looked at his father. Mr. Lambert hadn't moved. He was staring about him as though in shock. Virgil walked slowly to where James and Joshua lay on the ground in front of the door to the log house, the place where the porch would later be built.

Tears formed in his eyes when he saw the arrows protruding from their bodies. "Father in heaven—why?"

He squatted, feeling the muscles in his legs shake. Tears streamed down his cheeks as he leaned over one brother, then the other, checking for breath. His own breath came in shallow gasps as the realization hit him with the force of a rock smashing against his head.

"Pa! Stay where you are!"

Too late. Mr. Lambert was hovering over the bodies of his two sons, his face twisted in agony. "What—?"

He called out for his wife, his voice filled with anguish.

Virgil scanned the area. He saw his mother lying on a grassy mound about ten feet from the remains of the wagon. He rose, legs trembling, and crossed the ground slowly toward her. He knew.

A single arrow jutted from her chest. When he reached her, he heard labored breathing. It wasn't his mother's. Mr. Lambert stood behind him viewing his wife's dead body.

"O God! Why?" With long, choking sobs Mr. Lambert dropped to his knees beside the body. Then, to Virgil's dismay, he fell across her, draping her like a protective curtain.

"All my—fault," he muttered. "All my—fault."

Virgil stood there, frozen. The icy wind pierced him, causing him to shiver convulsively.

He recalled what the neighbors had said. "James looks just like the last owner—the one who killed an innocent Cheyenne warrior."

Faith had been right all along. They should have listened. He should have forced his father to listen! Then none of this would have happened.

Anger welled up inside him, mingling with the cold tormenting grief. *Faith, where are you? Charity?* Dashing to the wagon, Virgil searched around it. He circled the log home. Nothing. Had they been taken captive?

five

Evan saw Virgil dive for the rifle that lay a few feet from James's lifeless body. Dirt rose in a cloud around his face.

"It's me!" Evan called out. "Evan Dobson!"

He clenched his teeth until he saw Virgil release his grip on the gun and struggle to clear the grit from his mouth and eyes. Virgil spit, then stood up.

Evan studied Virgil's silhouette in the moonlit night. Then his gaze traveled to the blackened, skeletal remains of the wagon. The sight reminded him of a cemetery. He heard the older Lambert's wracked sobs.

Never in Evan's wildest imaginings would he have believed Colorado could present such devastation. This was *not* the "adventure" he had sought. His heart pounded harder when he thought of Faith.

Dismounting slowly, Evan approached Virgil where he stood by his dead brothers.

"What happened?" The question felt forced, but Evan didn't know any other way to ask it. His shoulders drooped, the weight of what he saw before him nearly unbearable.

At first Virgil remained quiet. When he spoke, his voice shook. "Indians."

He strode to a log where an arrow had lodged. He yanked at it, but it hung stubbornly in the wood. With one motion he bent the straight cherry shoot, snapping it near the base where the Alibates flint arrow point penetrated the wood.

Evan reached for it, taking it from Virgil's hands and turning it in the moonlight for a better view. He analyzed the straight groove which ran the entire length of the shaft. Rolling it over, he noted the zigzag channel that extended from the buzzard feathers to the flint apex. The point was still stuck in the log that had become part of the outer wall of the home.

"Cheyenne," he said.

"I know," said Virgil.

Evan was surprised that the young man knew. Evan himself had studied various tribes and their cultures, especially during his travels to Colorado. With his desire to become a surveyor, he hoped to help stake out the West for arriving settlers seeking new land and prosperity.

"Why?" he asked without thinking. He peered into the darkness for some sign of Faith, praying he wouldn't find her, yet needing to know where she was.

Again Virgil hesitated and swallowed hard. His tone was hollow.

"Neighbors. . ." He cleared his throat. "Some of our neighbors informed us that the Ruckers, the previous land owners, killed a Cheyenne brave in cold blood." He paused, shuddered.

Evan began to understand.

Virgil continued. "They said my brother," he peered down at James, "looked a lot like Mr. Rucker."

"And the Indians thought it was the same man," Evan finished, feeling his shoulders grow heavier still.

"Yes, so they came for their revenge." He drew in his breath.

Mr. Lambert's agonized wails had subsided. He moaned

only occasionally, as though life had slowly drained from him. In fact, both grief and exhaustion had worn him out.

Peering down at the two brothers, Evan saw they held hands, almost as if they had lain down and died beside each other, willingly. Had they made some sort of conscious choice? How? Why? He chose not to question Virgil about the strange sight. What bothered him most was that all the family members could be accounted for except Faith and Charity.

As if in answer to Evan's thoughts, Virgil whispered, "Where are my sisters?"

"We'll find them, Virg. Don't you worry," Evan said, placing a hand on the younger man's shoulder. *I promise we'll find them.*

"They took Faith and Charity," said Virgil, as if he hadn't heard Evan. "They took my sisters." His chest rose and fell sharply. He glanced at his father.

Evan squeezed Virgil's shoulder then released it. He knew it would be up to him to take charge. His temples throbbed at the possibility that Faith and Charity had been taken captive.

"I'll see to your family," Evan told him.

"Have a shovel around here someplace," Virgil mumbled, more to himself than to Evan.

"I'll find it." Evan surveyed the area.

"I'll look too," Virgil said, turning from the sight before him.

"Tell you what," Evan offered. "Why don't you take care of your pa—make sure he's okay? Take him off someplace, away from here." He looked into Virgil's tired eyes. "Can you do that for me?"

Virgil nodded. Then, with his chin nearly touching his chest, he made his way to his father's side. Gently he pulled the man up until he stood unsteadily on his feet.

Evan watched as Virgil led his father into the trees and down toward the river. Next he located the shovel and set to the task of laying Virgil's family to rest. Evan knew they could do nothing else until day broke. If they started the search now, in the dead of night, they might risk not finding the girls at all, a risk he wasn't willing to take.

The burying chore was completed just before the sun rose. As Evan viewed the round fullness of hot gold inching its way above the earth's rim, he said a silent prayer for the departed and one for the living.

Pondering his next move, his thoughts rested briefly on his own family. He knew that, having recently begun their travels, they would still be somewhere in Nebraska and wouldn't reach the Colorado border for some time.

He couldn't get word to them about the catastrophe—not that it would change their plans, but simply to warn them about the danger. The Dobsons were already committed to the journey. His own plans had certainly changed.

He would need help, Virgil's help, and Mr. Lambert's, if possible. As he considered his options, he heard footsteps. Turning quickly, Evan met Virgil's bloodshot gaze.

"Pa's sleepin'," he said, "down by the river."

"It's good he rests," said Evan, leaning on the iron handle of the shovel. Lifting the spade, he strode toward the unfinished log home, to draw Virgil away from the graves. He succeeded. Virgil fell into step beside him.

"Will you help me rescue my sisters?"

His question surprised Evan.

"Of course, I will." *You should know that.*

"How do we start?"

Evan leaned against the wall of the house. He was glad for Virgil and his father that the Indians hadn't burned the structure to the ground, like the wagon. Maybe they figured they'd better get away before they were discovered.

Drawing a cinnamon stick from his pocket, Evan bit off a small section. He slipped the rest back into his pocket and chewed on the piece in his mouth.

"We need to get going," he said after a moment. "But I'm worried about your father."

Just then Mr. Lambert staggered up the bank, rubbing the sleep from his eyes. Virgil walked the distance to his father and laced an arm around his shoulders, allowing the older man to lean on him.

Watching, Evan realized Mr. Lambert couldn't help them. The poor man was too disturbed and drained. *He'll have to stay here,* he decided. This presented a new problem.

He glanced over the house. It still lacked a small portion of the roof and needed more mud chinking and support logs. Most of the work had been accomplished. If they left him here, he would need appropriate shelter, or they would be gambling on the older man's life. He would speak with Virgil.

Already a plan had formed in his mind. When Virgil had his father seated on the ground where he could lean against the rough outer wall of the home, Evan approached him.

"We need to get started."

Virgil dropped his gaze. His chin quivered.

Evan fought to keep his emotions in check. Everything was still so fresh. But he understood that the longer they delayed, the harder the task of finding the girls would be.

"What about Pa?" Virgil's voice was hoarse.

"I need you to help me set a roof on this place, fast," Evan replied, sensing the younger man's concern. He had no idea how long they would be gone. "Your father will need shelter." He paused, studying Virgil's reaction.

"Where do we start?" Virgil gazed at the graves and then back at Evan.

"It won't take much time to get the rest of the roof up. The boards are ready. Then we'll get the mud in."

"And when we're finished?"

"We'll go after the girls and bring them home, God willing."

Evan forced himself to draw on a faith he knew he possessed yet had never needed to this extent. Until now.

⋅≈⋅

Faith and Charity slept in the shelter of the bush. During the night strange noises awakened them with a start, but they drifted back to sleep each time, weary from the trauma they had experienced.

At last the sun rose, waking Faith by its warmth. She decided to pay better attention to the direction in which they moved. She would study both smells and sounds. After saying a silent prayer for guidance, she woke her sister.

Faith struggled to her feet. "Ohhh," she groaned. Her body ached all over. She felt too stiff to move, but she knew she must.

Half stumbling, she distanced herself from the confines of the bushy shelter, while Charity stood up. Tilting her head back slightly, Faith allowed the sun to wash over her. She raised her hands toward the source of the warmth, her palms up. The heat hit her subtly from the east.

"Come," she told Charity. "We must start back."

"Which way?"

Faith took a step to the right. "North—I think. But I'll need your help. Okay?" Faith felt this was the right direction, though she didn't know why. Something inside seemed to be guiding her. *I am with you,* a voice seemed to say.

Gripping Faith by one shoulder, Charity led the way through brush and thicket, over rocks, around thistles and trees.

After covering some distance, Faith stopped, turning to her sister.

"Listen," she told her, "you must keep watch—make sure we're not spotted by anybody. Just in case they're in the area. Okay?"

She heard Charity draw in a weary breath.

"I'll watch," she replied.

Faith felt her sister's fear. Or was it anger?

"Virgil and Pa should be looking for us now," she said. "Have faith, Charity. We prayed. God answers prayer."

"Then what are we doing here? If God answers prayer, as you say?" Charity's tone was sharp.

"Keep your voice down," she whispered, touching her hand. Had Charity already lost sight of God? She had been so receptive when things were at their worst. Faith had hoped—

"Let's just keep moving," Charity said, "and you tell me which direction to go. *That* I can help with."

"Okay," Faith agreed. Her head throbbed. What horrors would they encounter that day? They'd had plenty already. How could they endure more?

෨

Just as Evan climbed onto the roof, he spotted riders in the distance.

"Virgil!" he called out, pointing. "Riders!"

As Virgil tried to get a better view, Evan kept his eyes on the strangers. His heart lurched when he realized one of the riders was an Indian.

"Get your pa to safety, Virg!" he warned. "And hand me up my rifle!"

"No need, Evan. They're our neighbors," he told him. "They settled a place last year not far from us, maybe five miles or so."

Evan struggled to regain composure. He wasn't taking chances. "Who's the Indian?" He noted that both were dressed in buckskin, but one was definitely native.

"She's half Ute, half Cheyenne, the wife of Barnabas Clayband. Her name is Sits Alone. They're good people, Evan. They'll help us."

Before Evan could respond, the riders were upon them. Both dismounted and walked over to Virgil, who greeted Barnabas with a handshake and Sits Alone with a tip of his hat. Evan watched from the rooftop.

"Thought we saw smoke in these parts," said Barnabas. "Decided to take a ride out."

"You saw smoke all right," Virgil said, pointing to the wagon.

Barnabas saw what remained of the family wagon. "Hmm," was his only response.

Then Virgil looked in the direction of the graves. The large man's gaze followed Virgil's. No sound came from Barnabas.

Sits Alone crossed to the side of the home where Evan had dropped the arrow. She bent over and picked it up, her mouth forming a frown.

Evan decided it was time to get down and meet Virgil's friends. Sliding to the lowest point on one of the eaves, he jumped to the ground, landing on his feet.

Virgil made the introductions.

"What ya got there, Woman?" Barnabas asked.

Her voice was soft, but her brown features were sharp. She handed the arrow to her husband. "Cheyenne."

Barnabas looked at Virgil, sadness etched in his own rough features.

Sits Alone stepped over to the graves, her beaded moccasins leaving footprints in the newly turned dirt. With hands at her sides, she studied each grave.

Evan spoke to the older man then. "Barnabas, we could sure use your help. We're putting a roof on this place so we can go get the girls."

"The girls?" Barnabas asked, his reddish-brown eyebrows arching over his eyes. "They got took?"

Sits Alone walked back and stood near them, listening.

"We think they were taken captive," Evan explained. "But it might take us awhile to find them, and we can't leave Mr. Lambert without shelter, not the way he is."

"Man's in a bad way, I reckon," agreed the older man. "Tell ya what—you two get your supplies and beat it outta here. I'll get the roof on. My woman can tend to Lambert."

Evan looked at Virgil with a questioning gaze.

Virgil nodded. "First I have to talk with my pa—tell him where I'm goin'."

"You do that," Evan agreed, "while I load our supplies."

Virgil walked over to his father who was still sitting where he had left him, staring straight ahead. Sits Alone followed and listened as Virgil explained everything to him.

Evan watched momentarily, wondering if the man even understood what his son tried to tell him. Turning, he made his way to the oxen, removed their yokes and their burdens, and hauled the items inside the house.

Barnabas was already hammering boards in place. Evan studied the man. He was of a different sort. Thick reddish brown hair hung long and unruly about his shoulders. He wore his fringed buckskin leggings and frock skin tight. Colorful quilled moccasins ran up his leg and stopped just below the knee. Evan guessed him to be around fifty and in excellent shape considering.

Later, Evan discovered that Barnabas was in this place before any of them arrived to settle the land. He made a living as a trapper, trading skins with the natives—mostly Ute, but some Cheyenne, Apache, and Kiowa—and was very successful because his wife spoke Ute, Cheyenne, and English. She also knew the general sign language used among the tribes of the area.

The trapper was as large as a grizzly bear and muscular. He seemed wild in one way and tame in another. *I wouldn't want you for an enemy,* Evan told himself, wondering if the man had any meanness in him.

Evan gathered hard tack, beans, and coffee in a *parfleche.* Sits Alone disappeared. When she returned, she handed another hide sack to Evan.

"*Pemmican,*" she told him, smiling. "Powdered buffalo meat, tallow, juice of berries."

Lifting the bag from her brown hands, Evan returned the

smile. "Thank you. I'm grateful."

"I will pray for you."

Evan hoped the astonishment he felt at her words didn't show on his face. "Thank you," he managed to say.

He wanted to ask her if she was a Christian but thought better of it. Obviously she was. What of Barnabas? Was he also? Neither fit Evan's image.

I'm wrong, of course, he thought. *But who am I to decide what a Christian looks and sounds like?*

He had to admit the contrast appealed to him in its own unique way. He suddenly realized he had been criticizing people all along. Subconsciously he had fit every Christian into an image he had developed sometime in childhood. Before he could ponder the revelation, Virgil came over to him.

"Ready?" Evan asked, thankful for the interruption. He was starting to feel uncomfortable with his thoughts.

Nodding, Virgil tightened Nat's saddle straps to make certain they hadn't loosened in the night. Because of the confusion, the young man had let his horse sleep in full gear. He climbed on the animal's back.

Evan lifted himself atop his mustang and glanced at Virgil, who touched the rim of his hat. The two young men rode away.

ஃ

"Charity, you're hurting my arm," Faith said as softly as she could.

"Quit whining, or I won't even touch you," Charity shot back. "Let's just see if you can find your way home on your own!"

Faith drew back at the sudden sharp words. They needed

to be together on this, to focus on the issue at hand. She felt Charity release her arm. Faith stopped at once.

"Besides," Charity continued, "I'm tired. I want to rest awhile."

Faith said nothing for a moment. She rubbed her hand across her forehead, wiping away the sweat. She was surprised Charity wasn't anxious to get home. Or was she? Maybe her stubbornness had reappeared, even in this dire situation, as though she couldn't help herself, no matter what. Faith kept her voice calm, despite her feelings.

"We can rest when the sun is straight up. Then I'll have to wait about an hour before we can go on—so I can feel a difference in the sun's heat, the way it warms me and where."

Charity was silent.

"Don't you want to get home?" Faith asked, hoping to entice Charity into continuing and wondering why she even had to do so. Faith wished desperately to be home. Being in the wild as they were, with no one but her sister, made her feel edgy. Besides she longed to touch Virgil and her father, *needed* to seek comfort in their arms.

"All right," Charity said at last. "But somebody better find us. And soon!"

"They will."

The two pressed on, with Charity guiding Faith around and over obstacles.

Again, though, self-doubt formed in Faith's heart. She tried to swallow it down as the hours passed. She told herself they both needed each other. Charity may have sight, but she had the sense of direction and other heightened senses. Neither could make it without the other. That was evident.

A new idea entered Faith's thoughts. *No one can make it alone. We all need the Father and each other.* Maybe she wasn't worthless after all. *Thank You, Lord, for helping me see this,* she prayed silently.

When she felt the sun warm her in equal proportions, Faith told Charity to rest. They sat in a patch of green bunch grass surrounded by white sweet clover.

"How much farther do you think we'll have to go?" Charity asked.

"A long way." Faith didn't want to be untruthful. Familiar feelings of inadequacy began to surface.

"I can't walk that far!" Charity said sharply.

Faith shrank inwardly from her sister. She felt helpless, and that feeling swelled inside her. Hadn't the Lord just spoken to her heart, telling her of her value? Her father and brother would need her more than ever now. Charity would not help them. Yet she could do nothing to remedy this! *Useless! That's what I am!*

Faith bit down on her trembling lower lip, as if to calm it. A tear traced a path down her sunburned cheek, and she promptly wiped it away. Charity mustn't see weakness in her. Faith knew she might use it for her own gain.

Faith wanted to cry out loud to her Father in heaven, and she would have, if Charity hadn't been nearby. Soon, however, weariness stole over her like a heavy blanket, and she lay down on the grass. But she didn't sleep.

❧

Evan clicked his tongue against the roof of his mouth

to gain Virgil's attention. If they spoke, it was only in whispers.

When Virgil glanced his way, Evan pointed to some blue sage. Something hung from one of its branches. He maneuvered his horse toward the brush, leaned down, and retrieved a piece of brown gingham. Virgil caught up to him, and Evan handed the cloth to him.

As Virgil took it, Evan saw his eyes light up. He knew immediately they were on the right trail.

"This belongs to Faith," Virgil stated, his tone low, but not enough to be a whisper.

"You sure?"

Virgil nodded, tossing the scrap of cloth to the ground. His lips formed two white lines of tension.

On impulse, Evan leaped off his horse and retrieved the cloth. He stuffed it into his pocket and mounted his horse again. The two rode on silently.

They followed a southeast trail of broken branches and tree limbs, even clumps of horse manure. The sun was straight overhead when Virgil prodded his horse on until he rode beside Evan.

"Can I ask you something?" Virgil asked.

"Sure."

"Why did you pick up that scrap from Faith's dress? Shouldn't you have left it there so we can find our way back?"

Caught off guard by the young man's sudden question, Evan was silent for a moment. "Why, I just thought we could add others to it along the way—be sure we're on the right trail. I'm sure we can find our way back without leaving it." He didn't know how convincing he sounded. He guessed Virgil was smart enough to hear what he wasn't saying.

"I think you're sweet on my sister Faith," he said.

Evan hesitated, apparently considering Virgil's words. It wasn't a question. Virgil had made a statement.

Memories of Virgil's protective nature surfaced. How might the brother of blind Faith feel about a man being attracted to her? The sister he would protect with his life?

"Yes," he answered, before he could convince himself otherwise, "I *am* drawn to Faith—always have been, as long as I can remember." A knot formed in his throat. He pulled a cinnamon stick out of his pocket, bit off a small piece, and put the rest back in safekeeping.

"How can you eat that stuff?" Virgil asked, his mouth puckering.

"Don't really know," Evan said quickly, glad for the change of subject. "Just always liked the spice. Most people don't, I guess." He turned to study Virgil's gaze. "You don't like cinnamon?"

"Sure, I like it." Virgil chuckled. "But I like it baked into things, sweetened, you know."

This was the first time Evan had heard Virgil laugh since the tragedy. It was a good sign. The young man would hurt for a spell, but he would bounce back. Virgil was a survivor.

"Yes, I know how that is," Evan said pleasantly. "I like it both ways, but mostly like this."

"Oh," Virgil said, shrugging his shoulders, "if that don't beat all."

Evan looked ahead, to each side, and then behind them. No sign of anyone else along the way.

Virgil cleared his throat. "What kind of feelings do you have for my sister?"

Virgil hadn't dropped the subject, after all. Evan realized the young man wanted answers. "I mean Faith no harm, Virg. I believe I love her, and I want to take care of her."

"She's blind," said Virgil. "She needs extra care, extra attention."

"I'm well aware of that. And I know I can give her what she needs. I *know* I can." *Shouldn't I be having this discussion with Faith's father?* he wondered. "I've felt this way a long time, Virgil."

"I don't know, Evan—I just don't know. Besides, what if she doesn't share the same feelings? She's never mentioned you."

Evan looked away, not wanting the young man to see the flicker of hurt in his eyes. Faith had been on his mind nonstop. He decided to change the subject.

"Is this Barnabas Clayband a Christian?"

Virgil nodded. "Yes, and so's his wife, Sits Alone. Good Christians, I might add. Why?"

"I don't know. I guess they don't seem the type."

A puzzled look crossed Virgil's face. Evan wondered if he had said the wrong thing.

"Who's the type, anyway? What's a Christian supposed to look like?"

At first, Evan didn't know how to respond. "I've never seen the likes," he finally said. "But I guess the Lord never set a dress code or skin color for becoming a Christian." He smiled broadly. "We just have to believe. Right?"

The revelation came easier than most. How many others,

Evan wondered, walked the earth with the same misconception?

Evan witnessed Virgil's smile of approval and relaxed. They traveled guardedly for several more hours, stopping once to drink water from a canteen and dip into the bag of *pemmican.*

Suddenly Evan heard a sound in some brush not far ahead. He raised his right hand in the air, warning Virgil to rein his horse to a stop. Virgil did so.

The sound came again. Somebody, or something, stirred ahead of them. Evan pulled his rifle out of the leather holster dangling from the horse's saddle, while Virgil readied his father's gun.

Both men dismounted and tied their horses to the thicket nearby. They crouched low, advancing slowly. When they were side by side, Evan caught Virgil's worried gaze.

He knew what the man was thinking. It could be the same band that attacked his family. How could the two of them fend off many warriors?

"Virg," he warned, "stay low and walk softly. Don't make any abrupt noises. Remember—we have the element of surprise on our side."

Virgil nodded, his gaze intent on the surroundings. Worry creased his forehead.

Evan drew in his breath and moved forward. He could feel his heart pounding in his throat.

six

A subtle wind delivered the scent to Faith in an instant. She rose to her knees from her lying position and spun around until she faced the direction from which the breeze floated.

"Somebody's here," she warned her sister. The warmth of body heat startled her as the girl pressed close to her. She heard Charity's frightened gasps and knew she must do something.

"We have to get to shelter," Faith whispered.

Lord, please don't let it be our attackers, she pleaded silently.

"Who can it be? I don't see anybody." Charity's voice cracked with fear.

"I can smell something," Faith whispered again.

At these words Charity stood and dashed to a nearby bush, leaving Faith alone in the open patch.

Rising to her feet unsteadily, Faith turned around, allowing her senses full rein. Her temples throbbed. Impulsively, she placed her palms against each side of her face as fear filled every part of her, settling in her fingertips and toes.

What is your name? a voice inside her seemed to ask.

"Faith," she whispered.

I protected you once, the voice seemed to say in gentle tones. *Do you remember, ye of little faith?*

Faith drew in breath after breath, as though trying to

draw her weakened faith into her lungs. At once, her heart eased its heavy pounding, and her hands dropped to her sides.

The familiar aroma of cinnamon both calmed and confused her. Evan. She would know his scent anywhere. Why was he here? His family wasn't due to arrive for another year.

And Virgil—she smelled him as well! Her chin trembled. Where was her father? She decided to call out, realizing they might have heard them and wouldn't know if they were friend or enemy.

"Virg? I'm here! Over here!"

Just then she heard what must be Charity scrambling from her hiding place—and fast footsteps as she ran in the opposite direction!

"Charity! Come back! It's okay!" Too late. Soon her sister would be out of hearing distance if she didn't stop running.

Faith cried out again. "Virgil!" Still no answer. She grew quiet, praying her brother would rescue her and wondering, with some vexation, why her sister had left her alone.

Evan heard Faith first. The sound of her voice, though laced with apprehension, caused his heart to leap. She was alive! *Thank You, Lord!* He stood up, shielding his eyes with his hand to find her.

"Faith! Is that you?" Virgil called out. He had come back after hearing his name shouted a second time.

At once Evan raised his hand sharply in the air to quiet Virgil. They must proceed with caution. The enemy could still be in the vicinity.

Virgil strode quickly over to Evan. "It's okay."

"No, we don't know—"

Virgil interrupted him, a smile chasing away the fear that had gripped his features. "Faith wouldn't do anything that might get us harmed. She would know if the enemy was near and would warn us. She would never call out."

Evan relaxed. Virgil was right. Faith reacted with calm resolve no matter what crisis she faced. Evan had witnessed this growing up. Charity was a different story.

Concern shadowed his eyes. *Where is Faith?* he wondered.

At last the men saw her standing in the grassy patch, still as a figurine on display in some fine clothing shop.

The two men dismounted, and Virgil hurried on ahead of Evan, reaching for the hand his sister held out and taking her in his arms.

Evan saw tears streaming down her cheeks, leaving glistening paths as they pushed through layers of dirt. He wanted to kiss away the drops of sadness and tell her he would take care of her from now on; never again would he let any harm come to her.

Faith wept openly now, as though a river of fear and grief had been dammed inside her heart and finally burst.

Evan realized Faith's journey had only begun. A mourning process would follow for her and the rest of the Lambert family. He told himself he would be there for her, if she would let him.

He gazed about the surrounding area in search of Faith's sister. Finally he noticed a head with mussed brown hair peering out from behind a tree. Knowing it was Charity, he waved in that direction, beckoning her to safety. Slowly, and hunched over slightly, the girl left her meager shelter and stepped into the clearing.

"Charity!" Virgil called out to his sister.

Tattered skirts flying, Charity ran to her brother and joined in the embrace. Her tears fell on the whiskers shadowing Virgil's cheeks and chin.

Evan stood apart, watching. How he wished to join the bittersweet reunion. But he knew he was an outsider.

His thoughts cleared, and he realized they must leave and make the journey home before nightfall. He walked quietly over to the little group.

"Virgil?"

Lifting his head, Virgil glanced at Evan. He understood at once and spoke to his sisters.

Charity headed for her brother's horse without even looking in Evan's direction, while her brother led Faith to Evan.

"Charity can ride with me," Virgil told him. "Can Faith ride with you?"

No words were necessary. *Of course Faith can ride with me,* he wanted to say. Instead he reached for Faith's hand. Gingerly he grasped her slender fingers and started to lift her onto his horse. But Faith didn't move.

Virgil and Charity were already on their horse, while Evan's snorted and pranced about. He pulled firmly on the reins to steady the animal and then drew his attention back to the beautiful young woman beside him. He sensed her fear and her inability to trust anyone now. She had lived through a trauma few survive. He understood. But he *needed* her to trust him. Completely.

Evan thought for a moment before an idea came to him. He remembered the barn dance and his first words to her. She had trusted him then. Hadn't she?

"May I have this dance? I'll be real careful."

He watched her chin quiver and felt her grip tighten around his fingers.

"Y–yes," she stammered, her whole body trembling. She stepped toward him.

"Now turn around, Faith."

She did.

In one movement he secured his hands at her waist and lifted her onto the saddle. She was light, and his motion seemed effortless. The mustang snorted.

Placing his left foot in the stirrup, Evan pushed off and swung his other leg over the top of his horse. When he was seated securely behind Faith, he fastened one strong arm around her slender torso. Then he clicked his tongue, set the reins across the animal's thick neck, and followed in Virgil's dusty wake.

&

Faith knew when they were close to home. New scents mingled with the old ones, some familiar, some oddly different, even disturbing. Grief flooded her heart. This was the place her family had taken their last breaths. Could she ever call it home again?

She recognized the first smells as belonging to family friends they had made when they first arrived. Apparently Barnabas Clayband and his wife, Sits Alone, were there.

"Where is my father?" Faith asked as Evan pulled the horse to a stop. The sun began to wane. There was a slight chill in the evening air.

"He's fine," Evan assured her. "Let's get you off this horse, and you can go to him."

Faith allowed herself to be lifted down. It felt good to be back on solid ground. How she wished to remove the tight lace-up boots that gripped her feet. Suddenly, hunger pangs pricked her stomach like tiny pine needles. But satisfying those would have to wait awhile longer, she told herself.

She knew the layout of the land they had settled. The scent of charred wood still hung thick in the air. Bile rose in her throat as the horrid memory stirred. She shoved it back with a deep breath and took several steps forward. One, two, three.

"Pa?" she called out.

Still walking, though somewhat unsteadily, Faith called out again. This time she heard a moan followed by a cough. She could smell him before he wrapped his muscular arms around her shoulders.

"My girl!" her father cried.

Faith pressed into the embrace, needing the escape it provided. For a moment she felt safe, as though everything was as it had been.

Then her father turned and called out. "Charity?"

He kept one hand gripped on Faith's shoulder as though afraid to lose her again. Charity ran to her father, and the three huddled close.

"The Lord be praised!" Mr. Lambert said, his voice husky with emotion. "I was so afraid you girls were gone."

It wasn't long before Virgil joined them. Tears flowed freely as Evan, Barnabas, and Sits Alone watched in silence.

॰੭

"We need your help," Virgil said to Charity.

Charity didn't care. It had become painfully obvious Evan was interested in her blind sister. Her shoulders squared with resentment. Everyone acted as if it was perfectly fine for it to be this way. But Charity had convinced herself that her sister could never give Evan what he needed in a woman. The girl was helpless as far as she was concerned. What could any man possibly see in a blind woman?

Some weeks had passed since their rescue, and everyday life had resumed, despite the tragic losses and times of grief. But Charity found herself wishing to leave the place. She felt like an outsider, now more than ever. Everybody else belonged in one way or another, even Faith. That fact was beyond Charity's comprehension. *Faith* was the strange one, being blind.

Then a thought occurred to her, bolstering her confidence in some bizarre fashion. Before she could carry the thought further, Virgil broke into her consciousness again. His presence grated on her nerves.

"Charity? Did you hear me?"

"I heard you all right!"

Virgil rubbed at the back of his neck. "I know you miss Ma, James, and Josh. I do too. But we must go on. They would want it that way."

Here we go again. Yes, she missed her mother and brothers, but she hadn't been dwelling on the loss. She had been too busy thinking about Evan, wanting him, and wanting him to want her. Charity was ready for marriage, anything that would take her away from this family, which gave her little to no attention, she thought. Faith this and Faith that—everything was always Faith. *It wasn't fair!*

At her silence Virgil stood up, set his Stetson back on his head, and stepped outside.

Sitting alone in the rocker by the hearth, Charity could think. She would *make* Evan notice her and would do whatever it took. She smiled to herself.

Faith had heard Virgil's pleading from where she was resting on a buffalo robe in front of the hearth across the room. Her heart ached for her sister. Maybe this was her own way of mourning. They had all suffered the loss of Mrs. Lambert and James and Joshua. But they had also kept busy, building a new life with the help of good friends who had chosen to stay awhile longer.

In a short time, they had completed the house and built beds with wooden platforms. A dining table fashioned from split logs and fastened together with wooden pegs sat by the window against the north wall of the home. Directly above the hearth hung a pair of deer antlers, six-point, and below that a Sharps rifle.

Faith was pleased Barnabas and Sits Alone had decided to stay. And Evan. He had stayed too.

But for some reason Faith found herself avoiding him. She sensed that the young man wanted her company—wanted more than that, in fact. But she couldn't comply. She wanted to, of course. After all, Evan was a sincere, kind, compassionate person—a man any woman would cherish in her life for an eternity. She had been drawn to him from the beginning, in a way she couldn't explain. But she could never give him what he needed. How could she be a wife when she couldn't do anything worthwhile? Evan needed a woman who could be by his side, working the fields and tending the livestock—someone who could cook and clean and contribute as a companion.

Believing she could be none of that, she chose not to respond favorably to Evan's advances. That would be wrong. Why ruin his life too?

Since their rescue Charity had refused to help the family in any way. And Mr. Lambert had been unable to do much at first. Finally, Barnabas had asked him to go out hunting for small game to help replenish the food supply. Faith knew the suggestion was more to help her father handle his grief by keeping him occupied than anything else.

Faith's thoughts returned to Charity. She knew it rankled the others that Charity would not contribute to getting the family back on their feet. Faith had done nothing to help the family either, and she felt ashamed. She hadn't even attempted menial tasks. But could she?

Faith stood up and brushed at her cream-colored muslin skirt to straighten it. Her mother had made this dress for her. She touched the strips of copper-colored lace that adorned the cuffs at each sleeve and around the hem. The thought of her mother made her chin tremble.

She walked to the hearth, taking small, yet accurate steps, all the while listening to the grinding of the chair as Charity rocked back and forth.

Without saying a word, Faith felt around until she found the water pail. She picked it up by the wire handle and turned in the direction of the door.

"What are you doing with that?" Charity asked sharply.

"We need water," Faith said, turning her face toward her sister.

Charity burst into laughter. Faith felt her determination wane, as usual. Her knees nearly buckled beneath her.

Lord, give me strength.

Charity stopped laughing. "You'll just fall into that river and be swept away, little blind mouse."

"Not if I'm careful." Her head was throbbing, and tears threatened. But she held them back.

"Oh, come on. We both know you can't do anything."

Faith fought to keep from giving up. "Guess I have to try, don't I?"

"What in the world for?" More laughter, muffled, as though Charity had covered her mouth with her hand.

"Our family needs our help," Faith told her sister. "I *want* to help. There must be something I can do." She hesitated. "They also need *you,* Charity. What's wrong, anyway?"

"Why, nothing, of course."

At that moment Evan entered the room. Faith caught his scent, and a strange, yet familiar comfort surged through her body. She wanted to shake herself but resisted.

"Faith?" Evan asked. "I–I need to talk with you. Privately."

Releasing the wire handle on the pail, Faith stepped in the direction where he stood. She sensed a desperate tone in his words.

"Can you come out on the porch?"

Faith allowed herself to be led outside. Sounds of hammers hitting nails and saws tearing into wood surrounded them.

"What is it, Evan?"

She recalled Charity's description of him: dark, muscular, sun-streaked hair, and green eyes. What did all that mean? Her own description must come through her hands and her other senses. How she longed to feather her fingers

over his face, lips, through his hair. How she dreamed—
This time she did shake herself. She hoped Evan hadn't
noticed.

She heard him lean against the log post that supported
part of the porch's roof.

Finally he spoke. "I'm not one for small talk—like to
get to the point." He paused.

Faith heard him draw in a deep breath. She did also;
then she spoke to break the uncomfortable silence.

"Whatever you must say, just say it."

"Good." He paused again. "I want you to know I'm in
love with you, Faith."

The words tumbled out, not as Evan had hoped to say
them. But he knew that if he didn't say them now, he
may not have the courage later. He also felt as though he
would burst if he didn't tell her.

Faith's eyes widened in shock, without her realizing it,
and she quickly pressed the palms of her hands to her face,
as was her habit when upset. Hearing Evan's chuckle, she
dropped them to her sides just as fast and grasped parts of
her skirt between her fingers.

And I love you too, Evan, she wanted to say but couldn't.

"I have been since the first day I laid eyes on you—in
school—in Nebraska," he continued. He hesitated, clearing his throat again. "I don't know how you feel toward
me, but I'm hoping that maybe you can come to have similar feelings. And if I need to slow down for you, I understand. Whatever it takes, Faith. I just want you to know I'd
like for you to be my wife someday."

Faith felt a strange sensation rush through her body, settling in her cheeks. She wanted to touch her face but

refrained. This was as close to a proposal as she'd ever encountered. At one time she had longed for this, believed in it. But now— Faith knew she must tell him the truth of her decision and then leave it at that.

"Evan," she began, "you're a good man, a worthy man. But you deserve somebody far better than me—"

"I don't want anyone else."

"Let me finish, Evan. Okay?"

He was silent.

Her heart beat faster. "I couldn't do much for you. I'm limited in several, if not all, areas. You need a companion who can work beside you—not somebody you'd have to take care of once your work day was over." Faith fought back the tears, almost believing she was giving away the opportunity of a lifetime, something of which she had only dreamed. And with a man she knew she already loved deeply. Then Charity came to mind.

"What about my sister, Evan? She would make a good wife, I'm sure." Faith knew Evan would never choose Charity. But she was searching for words, trying to diffuse the tension of the situation.

"I'm not in love with Charity. I'm in love with you."

And I'm in love with you too, Evan. She could never speak those words aloud, but they would haunt her forever. They were true; yet they could never be proved.

When he took her hands at the dance and led her safely into a wonderful new experience, she had loved him then—everything about him. She had heard people speak of "love at first sight." What about "at first sniff"? The humor was short lived, and tears stung her eyes. She swallowed several times before replying.

"Evan, don't you see? I can't do anything! I'm good for nothing. I'm blind!"

Evan touched Faith's trembling shoulder. Then he reached up and gently wiped a tear from her cheek. He let his finger linger there, stroking the wet spot and the supple skin beneath it.

She felt her body relax at his touch. He had a special way about him, a *very* special way. That was why she had fallen in love with him—for that and many other attributes that sight could never have secured for her as her blindness had. She knew this without doubt, but the realization offered no solace.

"That doesn't matter to me, Faith," Evan countered. "And you *can* do things. You've never been given the opportunity—that's all."

"Every time I've tried to do some menial task, I've failed. It would be so unfair to you. I'm sorry—so very sorry." Tears ran down her cheeks and hovered on her chin, waiting to break and fall, as she felt her heart had done already.

Unable to stand it any longer, Faith turned from him and hurried back into the house. He didn't follow.

Faith didn't know her sister had been listening to their conversation at a nearby window, a smile playing on her lips.

❧

Evan stood alone, a sense of hopelessness wrapping itself around him. The only thing he could hold onto was that she never said she *didn't* love him. *You must have some feelings for me, Faith.* She simply had no confidence, and he understood that. But he believed in her. She was an incredible woman in every way. He saw the potential even if everybody else was blind to it!

Just then Barnabas walked across the yard and stepped up on the porch, glancing at Evan. He pulled a large hunting knife from its scabbard dangling at his hip and, holding out the palm of one hand, began to dig the knife into his flesh.

"What's the problem?" Evan asked.

"Got a bear-sized splinter in my hand," replied Barnabas. "What's the problem with you?"

"What?"

"You seem worried—troubled, I reckon would be the better word."

Evan shifted his weight uncomfortably. What could this man possibly know about his troubles? And why was he concerned?

"Virgil told me you're sweet on Faith. I saw what happened on the porch—didn't hear nothin'—just watched, and the watchin' wasn't good."

At once Evan liked the man. He said what was on his mind. He respected that in a person. But could he be trusted with matters of the heart? Evan thought for a moment while Barnabas dug a long splinter out of his thick flesh. Who else could he talk to? Virgil was too protective. Charity thought only of herself. Mr. Lambert's spirit was still healing, and Sits Alone stayed busy cooking delicious stews and frying flat bread for the family.

"You can trust me, Boy," Barnabas nearly whispered, answering Evan's unspoken concerns.

I scarcely know you, Evan thought. *How can I trust you?* Yet something told him he could. Was it Virgil's earlier words about Barnabas being a Christian? The answer wasn't within easy reach, so Evan jumped in before he could talk himself out of it.

"It's Faith."

"Already know that much."

Feeling somewhat foolish, Evan reached up and massaged the cleft in his chin. "I'm in love with her, but she won't have me because she feels inadequate."

"Her blindness makes her feel that way."

A trickle of blood ran down the man's palm. He shook it off and returned the knife to its leather sheath.

"How can I help her feel *adequate?*"

Barnabas leaned against a corner post and stretched his hide-clad legs out in front of him. "It's somethin' the woman's got to do for herself."

His words caused Evan's heart to sink. "But I don't think she will. It's—like she's given up."

"This whole family's had more'n their share of misery," Barnabas said, taking in a deep breath, "Faith included. It's gonna take time."

"But what if she never comes around? What if she never discovers her own self-worth?"

"Not much you can do 'bout that, Boy."

Silence enveloped them at these last words. Evan felt a coldness grip him. "There has to be something I can do, Barnabas. Something!"

Standing up straight, Barnabas seemed to study Evan. "Maybe there is something you can do."

"What? Anything? Just tell me!"

"Won't be no easy task." He paused, rubbing at his red-whiskered chin. "You might have to deal with other family members as well."

Evan looked at Barnabas, waiting.

Turning aside, Barnabas spit into the dirt and then focused on Evan again. "You can teach her how to do things. Might be things she doesn't want to attempt. And you could have problems with the little brother over it." He glanced at Virgil.

"What do you mean?"

"Virgil's very protective where Faith's concerned. You've maybe noticed that. More so since losing his twin. He wants to do everything for the girl, not knowing any better, of course."

A plan began to take shape in Evan's mind.

Before Evan could give voice to his thoughts, Barnabas turned and rested both palms on the porch railing. "If you force Faith to do tasks, even if she gets hurt a little, her insecurities may deepen. I reckon that wouldn't last long when, and if, she took an understanding to what was happening. 'Sides, it'd be for her own good either way. Faith needs to be able to live, really live. She probably won't get the chance otherwise."

Evan stepped off the porch and looked up at the sun, then back at Barnabas. "What if she resents me and then won't have anything to do with me—ever."

"That's a risk you'd have to be willing to take, I s'pose. At least you would have helped her find some amount of freedom. You'd have given her a gift. But are you willing to take that chance?"

Kicking at a clod of dirt, Evan pondered his options. If he did nothing, he risked losing her. And if he did something, he also risked losing her. Yet the latter would at least be of help to Faith. He wanted her happiness more than anything, and he would put himself aside, if need be, to help her find that happiness.

Barnabas's voice broke into his thoughts. "You might end up with a very angry woman on your hands."

"Why?"

Barnabas winked but didn't smile. "You'll learn the answer to that soon enough." He stepped off the porch and disappeared from view to resume his tasks.

At that moment Faith managed to steer herself out the front door with the water pail in her hand. The bucket clanked against the door jamb, the porch railing, and the steps as Faith worked her way down them. He knew she knew he was there. Apparently she didn't care.

Evan watched as Faith fairly stumbled across the meadow, heading for the thick trees and brush that lined a swollen Rio Grande river. Clenching his teeth, Evan followed, his heart pounding in his chest. He couldn't see Faith any longer. He was running before he heard the scream.

seven

Charity smiled at the sound of the cry. She hurried out the front door and followed the others down to the river, all arriving at the river at once. Mr. Lambert was still out hunting wild game and probably did not hear his daughter's screams.

She was fine—wet, but fine. *I warned you, little blind mouse.* A grin twisted Charity's lips. Obviously, her ill-equipped sister had stumbled and fallen into the water, but only at the edge.

In her panic to scamper out, Faith had let the bucket float downstream. It was clinging to some low-lying bushes hanging over the water. It had caught on one of the branches by its wire handle and now bobbed, almost comically, in the river's currents.

Charity's thoughts were interrupted when she caught sight of her brother dashing to Faith's rescue. But Evan intercepted Virgil, halting his efforts. She knew what he was doing, having heard part of the conversation on the porch.

Evan wanted them to give Faith the chance to do things herself. But Charity knew Faith would fail at anything she attempted. She knew her sister, had lived with her too long.

Charity was glad when Virgil shoved past the man.

"Hey! What are you doin'?" Virgil shouted. "Get out of my way!"

Charity glanced over at Barnabas and Sits Alone, noticed they stood still, watching, and then looked back at Evan.

"She's fine, Virg. Leave her be," Evan said.

Virgil was scrambling to the river's edge where Faith stood crying, soaked and muddy on her right side.

Charity knew her sister's tears were due to frustration.

"Virgil?" Faith asked, looking toward the sound.

Virgil reached for his sister, supporting her around the shoulders, and started for the log home.

"Why'd you do that, Faith? You could have been swept away—drowned!"

Faith didn't answer. Sits Alone followed them inside.

Barnabas was now standing beside Evan near the banks of the river. Evan raked a hand through his hair. From where she stood, Charity could hear the older man's words.

"Give it time, Boy," the big man explained. "Give her a chance to prove herself."

Inside the house Virgil asked the Indian woman, "Can you help her change out of her wet things?"

Sits Alone nodded. The Indian woman stepped to Faith's side where she took over Virgil's position, securing an arm around her back and shoulders.

Virgil left then, closing the door behind him, and stood on the porch for several minutes, watching Evan and Barnabas talking.

Charity saw that the two men were still by the river and decided her opportunity was at hand. Quickly she made her way to the river bank, to the place where the bucket was caught on the branches, lifting her skirts to her ankles to avoid stumbling. She could no longer hear the men talking and hoped they were watching her.

Lifting her skirts to her knees now, Charity positioned herself astride the thickest branch, as though mounting a horse, and scooted along the limb. She worked her way out until her feet dangled over the water and the limb bowed with her weight.

She felt a slight panic when she looked down at the swirling, rushing waters, swollen from the heavy spring rains. But thoughts of Evan's green eyes watching goaded her on. She would show him what *she* could accomplish. Where Faith failed, *she* would succeed.

As she stretched an arm toward the bucket, a lock of her long brown hair blew across her eyes. Charity shouldered it away, needing to use one hand to stabilize herself and the other to retrieve the pail. Reaching out as far as she could, she managed to grip the wire handle, and finally, by pulling hard, the bucket was free.

Charity pulled it to her and, holding it in one hand, inched her way backward until she reached the base of the branch. As she stood up, she glanced toward the men, noticing that Virgil was now with Evan and Barnabas. No one spoke. All eyes were on her—as she had hoped. Smiling at the men, the girl sauntered over to the river's edge near them, filled the bucket with water, turned, and walked to the house, where she disappeared inside.

Virgil turned his attention on Evan. Before he could say a word, Barnabas spoke up.

"We'll be needing to get word to Fort Garland."

Evan felt relieved. He understood Virgil's anger but wasn't sure Virgil comprehended what he was doing concerning Faith. He was glad for the interruption.

"What?" Virgil asked, focusing on the big man.

"Word of the attack—to the fort."

"Oh—yes." Virgil glanced at Evan and then back at Barnabas. "Will you be going?"

Barnabas fingered the knife in its leather holder on his hip. "Probably best for you to go, Virg." He paused, as though anticipating a negative reaction. "I'll stay here—take care of these folks."

"Why can't Evan go?" Now Virgil faced Evan.

The thought of leaving brought a lump in Evan's throat. He reached into his pocket and pulled out a piece of cinnamon. Popping it into his mouth, he chewed it slowly. The chewing or the cinnamon, or both, calmed him—he wasn't sure which. He studied Virgil's cold, gray gaze. He knew the younger man would as soon have him gone so he wouldn't interrupt the natural flow of life as Virgil saw it. But the flow had kinks in it that Virgil apparently couldn't, or wouldn't, see.

To Evan, this flow meant the involvement of every able body, including Faith. Yes, she was limited because of her blindness, but who, including Faith herself, knew what she could accomplish? And she would never learn without the opportunity. The lessons should have been started long ago; now there would be plenty to make up for in a shorter amount of time. But Evan believed in Faith and in her abilities.

Of course, he understood Virgil's desire to protect her. That was natural. But the family unwittingly had sheltered Faith. She felt helpless and frustrated and therefore inadequate. Evan felt like the villain because of wanting Faith to grow through learning her strengths and becoming independent.

Evan knew he would take Faith either way, helpless or not. He loved her in spite of her limitations. Her happiness, though, was more important to him than his own. A sense of dread filled him, however, as he realized he might lose her. Was it worth the risk? Yes. The answer required no thought. He wanted her to be happy either way—though, of course he preferred she find it with him.

Evan shrugged off his thoughts and tried to catch, mid-sentence, what Barnabas was saying.

". . . better with a gun." The older man was scanning the landscape. "We can protect your family if there's another attack. You know that, Virgil." He hesitated, glancing from one young man to the other. "I'm not saying you're not a good shot—just that we've had more experience."

Virgil shifted his weight to his other foot and shoved his hands into the pockets of his jeans. "I can protect them too," he muttered gruffly.

Evan understood what Barnabas was up to. Yes, the army needed to be notified. The man was trying to get Virgil to go so Evan could do what he could for Faith's benefit. Nothing would be accomplished if Virgil kept interrupting the process. He felt for Virgil, knowing the brother was miserable. Having just lost his family, it would be difficult to leave the remaining loved ones in somebody else's hands.

When Virgil raised no more objections, Evan felt a sense of relief but said nothing.

The older man put his rough hand on Virgil's shoulder and squeezed it.

"I'll get my gear together," Virgil said, his mouth forming a hard, straight line. He strode over to his horse which was grazing on the plush bunch grass.

Evan looked at Barnabas. His opinion of the man had changed. He trusted him now and believed him to be a solid Christian, more so than most he'd met. He didn't preach, was matter-of-fact, and had no surprises or hidden motives. He was what he appeared to be.

"Thanks," he told the man.

"No need. I don't envy you and what you're getting into," Barnabas replied. "Wish I could be of more help, but I can't. This is something you alone must do."

Nodding, Evan glanced toward the homestead. "Maybe I'll go up and talk with Faith. See if she's sore at me."

"Reckon she is," Barnabas chuckled.

Evan smiled at that. Then he turned, ran his fingers through his wavy brown hair, and walked up the knoll to the home.

ða

Clothed in a fresh pale pink linen dress, Faith rested on the thick buffalo robe in front of the hearth. Shivering, she drew the colorful, star-burst quilt around her shoulders. She felt as cold inside as out. She had tried to accomplish a simple task and failed.

Sits Alone stoked the fire and added more wood until a good blaze warmed the dwelling. Faith turned her face toward the heat, relishing the feeling.

Faith sensed when Sits Alone knelt before her. She smelled like the outdoors, in a pure way. Faith knew the woman had washed her hair earlier with yucca suds while bathing downstream where the banks squeezed together. Boulders protruded at various points, allowing the water to flow more gently in certain areas.

"Warm now?"

"Yes, thank you." Fresh tears rose in her eyes. She swallowed hard, as though gulping emotions back.

"You did good," Sits Alone said.

"What?" Faith asked, incredulous.

"You tried. That is good," the woman told her.

"I fell in the water," Faith said. "I couldn't even get water for my family!"

She heard a laugh from across the room. *Why must Charity be like this?* she wondered. Then Faith heard a noise as Sits Alone clicked her tongue on the roof of her mouth. Footsteps sounded across the floor, and the door swung open with a creaking noise. It didn't close. Faith assumed Charity had gone out and left it open, upset at Sits Alone's response to her laugh. Faith had no idea Evan stood just outside, listening.

"You must do it again and again," Sits Alone urged the girl, her tone soothing.

"I couldn't even do it the first time," Faith argued. "I'm completely helpless. I can never be of any good to my family, much less a husband." She placed her palms against both sides of her face in an effort to cool her cheeks.

Sits Alone grasped Faith's pale hands gently with her own brown ones and lifted them away from her face, settling them in Faith's lap.

Faith drew in a deep breath. A tear trickled down her face, hovered on her trembling chin, then dropped, disappearing into the bodice of her dress.

"Let somebody help you, as you learn," said Sits Alone.

"No," Faith said firmly. "That's been the problem. Every

time somebody tries to help me, they end up doing the task for me. I'm just there for company."

Sits Alone spoke again, with compassion and wisdom in her words. "We are all taught from birth how to walk, to talk, and much more. If we are taught in the right way, we *will* learn. No matter how old we are, we can still learn."

It had never occurred to Faith that she had been taught in the wrong manner. In fact, it seemed impossible to her. Her family had done the best they could. After all, they'd had poor Joshua to deal with too. No, she told herself, they had taught her right. Other questions haunted her.

"Why would God do this to me?" she asked, not expecting an answer. She knew there was no answer to that question.

Sits Alone thought otherwise. "God gives us everything we need to survive the journey set before us. You have many gifts, Faith, for you to use as you walk along your path."

Faith ignored the woman's words and asked another question.

"You believe in God, right?"

"Yes, I was converted to Christianity five years ago. My Cheyenne mother converted before that."

Swallowing hard, Faith continued. "When—when I was lost—I mean, when my sister and I were lost, after being captured, I prayed. Then a Scripture verse came to mind, one I'd heard before. I believed that verse had special meaning for me and wanted to believe that with all my heart."

"I can't read the white man's language, but my husband reads the Bible to me often. What Scripture verse did Father send to you?"

"Isaiah 42:16."

"Do you remember it? Can you speak it?"

Faith didn't hesitate. " 'And I will bring the blind by a way that they knew not; I will lead them in paths that they have not known: I will make darkness light before them, and crooked things straight. These things will I do unto them, and not forsake them.' " Fresh tears flowed down Faith's cheeks as she completed the verse.

"Tell me what you feel about this," Sits Alone asked.

Clasping her hands in her lap, Faith answered, "I believe the Lord has forsaken me."

"No," the Indian woman said quickly. "He has not. You feel that way."

"Then what does it mean?"

"Only you can answer that," Sits Alone told her. "It was given to you. Now you must understand what it means."

"Crooked. What does crooked mean? Or straight? And how can my darkness become light when I live continually in darkness as I understand it? I don't even know what light means."

"You must think on these things," Sits Alone said. "And I will try to help you understand. But you must also accept the gifts God has already given you first. You must believe in yourself before understanding can enter your heart."

"Believe in myself? What gifts do I have that could help anybody?" she asked, with sudden scorn in her voice.

"You can smell as well as any animal, and your hearing is strong. Faith, it is as though you are 'one with the wind.' "

"What do you mean—'one with the wind'?"

."The wind brings things to you; shares a part of herself with you. You know of danger before anybody else. Virgil explained how you tried to warn your family before the attack. You knew. The wind is your sister."

"What does the wind look like?" Faith lifted her hands in the air and let them fall back into her lap. "Describe it to me."

Sits Alone was silent. Faith knew she was thinking. She waited.

"One can't see the wind—only feel it," she said after a moment. "Just like we can't *see* God, the Great Power, *Heammawihio,* but we can *feel* His presence."

Faith understood about the Father and how she felt with Him in her heart. "You can't *see* the wind?" Faith's voice was filled with amazement. "I thought people with sight could see everything in the world."

"No," Sits Alone assured her, "we cannot see everything. Like the wind. We can't see it any more than you. Maybe you, Faith, actually 'see' the wind better. Because you are 'one with the wind.' This is a gift. I would love to call the wind my sister, but I cannot."

At first, Faith didn't understand. It baffled her that nobody could *see* the wind any better than she, and yet she had a special connection to the wind, through her other heightened senses. She tried to let the words of Sits Alone sink in, tried to understand their meaning, but it was difficult. Her curiosity piqued, Faith continued.

"And what does crooked mean? Or straight?"

Again, a long silence enveloped them. She heard the woman rise from her position on the floor across from Faith.

"I will return," was all she said.

Faith sat alone, running her fingers through the rough fur of the buffalo robe. For a moment she caught the scent of cinnamon; then it was gone. Faith wondered if Evan was nearby. Maybe he had walked past on the way to his tasks.

The thought of Evan caused fresh, warm tears to brim her lashes. She wiped at her eyes with the hem of her skirt.

Yes, Faith knew she wanted to be his wife. They had grown up together, but it was during the dance in Nebraska when she had fallen in love with him and didn't realize it. She knew she couldn't stop thinking about the man after that. Flowers took on a new significance, their petals so full of softness. In that one moment she thought of a husband and children. A life full of love and hope. But Charity always reminded her of how foolish she was to dream in such a way. And how correct Charity was!

" 'And I will bring the blind by a way that they knew not. . .and not forsake them.' " The words rolled off her tongue almost of their own volition.

A new thought entered her mind. "Lord, have You put me on a path I haven't known?" Just as she asked the question out loud, Sits Alone reentered the dwelling, leaving the door open.

Faith felt the woman's body warmth and smelled her pleasant scent again as she sat down across from her.

"Let me have your hand," Sits Alone said. "I want you to touch something."

Stretching her fingers toward the woman, Faith allowed her to take her hand. Sits Alone guided Faith's fingertips gently over the item she held.

"This is a Cheyenne arrow," the Indian woman explained.

Faith jerked her hand away at once.

"No, no," Sits Alone said softly. "It will teach you something. The fault belongs with the people who took their lives, not with the weapon." She paused. "Give me your hand."

Still Faith cradled it in her other hand as if to protect it. She hated arrows. And, yes, arrows *had* killed her loved ones and set fire to the family wagon.

"This arrow never touched your clan," Sits Alone assured her, "only lodged in the wall. It is not dangerous."

Reaching out, she grasped Faith's fingers again, holding tightly. "The fault belongs with those who killed your clan members," she repeated. "Let me teach you."

Her hand trembled, but as Sits Alone held it, the trembling eased.

Beneath her sensitive fingertips, Faith felt a groove that shifted direction several times.

"*Tsis tsis tas,* my people, make their arrows with a crooked groove that runs along the cherry shoot shaft," she explained, letting Faith feather her fingers back and forth over the groove.

"This is what 'crooked' means. It is not straight."

"And straight?" Faith asked.

Sits Alone turned the arrow over, exposing the other rounded side. Taking Faith's fingers, she guided them along the straight groove.

"And this means straight."

Taking the broken arrow out of Sits Alone's hands,

Faith brought it into her lap. She turned the arrow over and over, letting her fingers travel down the shaft—first one side, then the other. "Straight—crooked—straight."

"Yes," said Sits Alone, sighing audibly. "Does that help you understand your verse now?"

"I'm not sure."

"Think on this—if you were walking along either of these grooves, which would be the easier path to travel?"

"This one," Faith replied, running her fingers along the straight channel. "He will make 'crooked things straight.' "

"Because straight is the better way, the right way."

"And light? What is light?" Faith was eager to learn, to understand. She thanked the Father silently for bringing Sits Alone to them. She and Barnabas were helping the Lambert family immeasurably.

She heard Sits Alone shift to a more comfortable position. Faith moved her own legs out in front of her, wriggling her toes to increase the circulation.

"I'm not sure how to explain light to you," the older woman said, clearing her throat. "Tell me what you see, Faith."

"I don't see," came Faith's fast reply. "You know that."

Sits Alone was undaunted. "Do you see blackness?"

"I see nothing. I don't even know what 'blackness' means. I see nothing, as I know 'nothing' to mean."

"Light would be the opposite of that 'nothing' you speak of."

Faith tried to grasp the meaning of her words. "I–I don't understand."

"What you are seeing is darkness. Do you understand?"

"Yes, I guess so."

"The other side of darkness is light."

"Does light mean 'seeing'?" Grasping this concept was difficult for Faith.

"Not necessarily, because most people can still see in the dark."

"Oh."

"If you'll think about the words of that verse, you might understand them better."

Faith decided to tell her a part she *did* understand.

"Charity is blind, like me, but in a different way. One time my brother, Virgil, told me that both of us had the same instruction growing up—but Charity only heard the words, while I *listened*. The teachings were darkness to her. But I *listened*," she emphasized.

"With your heart."

"Yes. It is the same way with blindness. My sister is blind by way of her spirit. I am blind by way of the flesh."

"This is true." Sits Alone paused. "Though Charity can see in light, she walks in darkness. You see only darkness; yet you walk in light. Charity is blind to the thing most valuable, our salvation. Yet you *see* this!"

Faith felt warm inside. She now knew what straight and crooked meant, and darkness and light, in her own way. But the thought of Charity chased the warm feelings away.

"Will my sister be okay? I worry for her."

"She should learn as you learn—as we all must learn. But if her heart is not open, understanding cannot pass into it. Your heart is open, Faith, and understanding will enter. It will bring light into your spirit." She leaned toward Faith.

"You must let yourself learn things, and you must let others teach you, as we all are taught."

Again Faith caught the scent of cinnamon. She chose to ignore it. "Will—will you teach me, Sits Alone? Will you be my teacher?"

"I will teach you all I know, if, in return, you will teach me."

"What can I possibly teach you—"

"That's for sure!" Charity broke in. She was standing in the doorway, her hands on her hips. "Faith can't teach anybody *anything,* because she doesn't know anything!"

A gentle breeze came off Sits Alone as she stood to face Faith's sister. "We are all students, and we are all teachers," she said, her voice steady.

"What are you going to teach Faith anyway?" Charity asked sharply, crossing the room to the hearth.

Faith wanted to cower but refused to do so. If her sister meant to slap her it would happen whether she cowered or not. She would not keep giving in to Charity's insolence and abuse.

The sound of Sits Alone's voice calmed her racing heart. "I will teach your sister all I know."

"Which isn't much," came Charity's caustic reply.

Sits Alone chose not to respond. Charity continued, apparently feeling empowered by the silence.

"It was people like you who killed my family—Indians, savages!"

"Charity! Stop!" Faith could remain silent no longer.

"Well, it's true," her sister answered. "And I can't believe you would want to learn *anything* from somebody who did this to our family!"

Faith stood up, grasping her skirts in her fists. "Sits Alone did nothing to our family. All she's tried to do is help us."

Charity's words came fast. "Cheyenne blood runs through her veins. She's the same as they are. And so is Barnabas Clayband for marrying her in the first place!"

"You can't judge a person on the actions of others, no matter what blood runs through their veins!" Faith trembled in anger. Her heart beat faster.

Sits Alone spoke before Faith could respond. "I have killed no one. I have no wish to do so. Nor am I accepted among my own people, the Cheyenne, or among the Ute. My mother named me Sits Alone because I don't fit into any culture, not even the white eyes' world."

Tears filled Faith's eyes and rolled down her red cheeks. She wiped them off as quickly as they fell. Without sight she saw no skin color—therefore, knew no prejudice. At that moment she wished everyone, including her sister, could be blind, if only for awhile. They would learn there is a unity in the human race, with a connection through something as simple as scent. For the first time Faith realized her blindness had taught her something. She could see everyone as equal, as God saw them.

"You don't belong here, Sits Alone," Faith heard her sister say. "And I don't want you teaching *my* sister anything. Do you understand that?"

"I'll be the judge of that, Charity!" Faith exclaimed, her voice trembling with rage. "Sits Alone will be a wonderful teacher, and I'm ready. More than ready!" She turned to face where she knew the Indian woman stood.

"When do we begin?"

"When you choose."

Both women felt the floor boards vibrate as Charity marched across the room and through the doorway. At that moment Evan stepped up, caught the door to keep it from slamming, and entered the room.

The scent of cinnamon passed Faith's nose, and she bristled, not sure if she could handle much more.

Evan spoke first. "I heard what happened. It seems Charity sneaked in. Everybody okay?"

"Yes, fine, thank you," Faith said, with no inflection in her tone.

Hearing a noise outside, Sits Alone left the room abruptly. Evan and Faith stood alone.

"Are you angry with me, Faith?"

"Indeed, I am, Mr. Dobson," she said sternly, her thoughts still on the encounter with her sister. She drew in a breath, trying to adjust herself to the new situation. "I heard you try to stop my brother from helping me."

"I did it for your own good, Faith."

"How can you say that? I could have drowned."

"If I'd thought you were about to drown, I would have been the first one in the river to rescue you. I promise you that."

A coldness settled in Faith's chest then moved through her body to her fingers and toes, making them feel like ice.

"Well, don't do me any favors—from here on out!" The words ached all the way to her temples. She was confused, and her emotions were getting the best of her.

Just then Sits Alone appeared at the door. "Your father's back from the hunt. We have work to do."

Faith stepped toward the sound of her voice until she was outside, leaving Evan alone and disoriented.

eight

Grady Lambert unloaded five rabbits and a doe, gutted and skinned, from his horse. He had carried the doe in front of him, its limp body slung over the saddle. The rabbits he'd strung together with rope and hauled across his shoulders.

Faith wrinkled her nose at the scent of death but knew God had provided these creatures for nourishment and warmth. She stood very still, wondering what Sits Alone might have her do. *Will I fail again?* she asked herself, her heart quickening its pace.

She felt the gentle touch of the Indian woman as she whispered, "Come."

Without support Faith had to follow by sound and scent. Each time she stumbled, she righted herself and proceeded. She was determined to continue. *I will make darkness light before them, and crooked things straight.* She imagined herself walking along a straight groove, like the one she felt on the arrow's shaft, and to her surprise the way became easier. She stumbled less often, and her trembling eased.

Something soft touched her hand as Sits Alone gave her a rabbit's fur. Faith grasped it in both hands and stood waiting.

"I want you to hold this *vohkoohe* until I return," the woman said. "Touch it until you know it well."

Dropping down into the plush grass near the river, Faith

did as she was told. She manipulated the soft, slack pelt, touching every fiber, every curve, even the fleshy substance that clung to its underside.

After some minutes Sits Alone returned.

"Ohhh—what's that smell?" Faith asked, turning toward the unpleasant odor that had just reached her nose.

"Heated brains, liver, grease, soapweed, and water—all mixed together," the woman replied lightly.

Faith suddenly felt sick to her stomach. "Whatever are you using *brains* for?"

She had never heard of such a thing. The smell was almost too much for her. It wasn't the overpowering odor a skunk would emit. Rather this smell had a subtle, yet pungent, quality that seemed even more disagreeable.

Sits Alone chuckled softly. "You want to learn?"

"Yes."

"Lay your skin out flat, flesh side facing the sun."

Using her hands Faith made sure no part of the skin was folded or creased, that it was completely exposed.

"The brain mixture is for tanning the hide. But first we must scrape." She paused. "Here—take this."

Faith reached toward Sits Alone's voice and felt the hard object placed against her palm. She was grateful this woman had become her friend and now her teacher.

"The 'flesher' is made of buffalo bone," she explained. "Use it against the flesh to begin removing it. Be careful that you don't break through the skin."

Before starting, Faith studied the object with her hands. The cannon bone of a buffalo had been cut off diagonally from above, angling downward toward the distal end. A hole had been drilled through the proximal end and a thong of buffalo hide passed through that hole, through which, then, the hand was passed.

Slowly, methodically, Faith scraped. She used her fingertips to feel a specific soft section before running the sharp bone across it. As she progressed, she had to toss wasted fat and membrane to the side. Faith heard Sits Alone working another hide in front of her.

"That is good," her teacher praised her. "You do well."

"What next?" Faith asked, anxious for a finished product.

She had never created anything. Her mother had sewn garments for the family, even knitted shawls and house slippers, but Faith had never been part of those activities, though she had wanted to.

Some items were placed beside her. Touching them, she realized one was a pottery bowl filled with warm water; the other, a flat smooth stone. She lifted the stone and set the flesher near her right leg so she wouldn't lose it.

Sits Alone rose from her work and stepped behind Faith who knelt on the ground. "Take the stone, dip it in the water, and rub it on the skin to dampen it and remove fat and membrane you might have missed."

At first, Faith struggled, trying to adjust to the new tool in her hand. With her other hand she held the skin down and ran the stone back and forth and to each side.

Time passed quickly. Faith began to hear the sounds of the night and felt a slight chill mingle with the air as the sun dipped below the horizon. Then she smelled cinnamon, and something twitched in her stomach.

Tears wanted to surface, but she held them back. Was Evan watching? she wondered. She had to believe she was doing the right thing. He could have anybody. And he deserved happiness. Faith wanted Evan's happiness more than anything, even if she couldn't have him. When one truly loved, the other person's happiness was foremost in the heart, she believed.

Still, a spark of hope kindled inside her. If she could learn, as Sits Alone asserted, maybe she could become the wife Evan needed. If she became self-reliant, then she would feel good about marriage and a family.

Sits Alone set a bucket on the ground beside her, interrupting her thoughts.

"Here," she explained. "It's still warm. Rub this mixture into the skin using the stone in your hand."

Swallowing hard, Faith dipped into the smelly, mushy mixture and spread it over the rabbit hide using the stone to work it in.

"Don't miss a spot," said Sits Alone. "We must be quick, though, as our light is almost gone."

At this, Faith laughed out loud. "I can do this through the night," she said between her giggles. "*You* may need light to see, but I don't."

"Ya!" Sits Alone joined in the laughter. "See—you have something most people do not have. You *are* blessed."

Blessed? Was she? Faith dismissed the idea. She couldn't see it yet.

Evan moved downwind of Faith and Sits Alone. He could tell she was unaware of his presence as she worked. He delighted in her laughter, remembering how she had laughed when they were children. A delicate yet full sound left her perfectly formed lips and seemed to enter his heart in a place reserved especially for her. Quickly, though, he felt a heaviness in his chest.

Evan had never intended to make her angry. Barnabas was right about what Faith needed. He thanked his Father that Sits Alone was taking such pains with the woman he loved. But would Faith forgive him? Had she lost her ability to trust him now? A silent prayer took shape in his mind.

"Why, Evan! What are you doing out here?" The voice jarred him from his thoughts.

The young man turned and saw Charity holding a tin plate in her hands. He smelled the food, and at once his mouth watered. He hadn't realized how hungry he was.

"I made this myself," she said, pushing the plate toward him.

Out of politeness, he accepted the meal and had taken the first bite when she spoke again.

"Bet you're plumb tired of eating that Injun's food. I certainly know I am."

Evan dropped the spoon at once and handed the plate back to the girl. His brows gathered in a frown.

"I don't take kindly to what you just said," he said, with sternness in his voice.

"Oh! But it's true."

"No, it isn't! Barnabas and Sits Alone have stayed on to help your family without being asked. They're good people, Charity, and you know it."

Suddenly he remembered how quick he had been to judge the Claybands and knew he mustn't be too harsh with her.

"How can you say that?" Charity's deep brown eyes widened. "It was people like that squaw who killed my family!"

Evan chose his words carefully. He realized she could be talking from her grief and not thinking clearly.

"I'm sorry about your family, Charity, but Sits Alone did *not* kill them. She had no part in it. She has only tried to help."

Charity tilted her head, shaking her silky brown hair and causing it to rest along her back. "I wish you would eat something, Evan," she said, ignoring his remarks. "I'm worried about you."

For a moment his heart softened. Was Charity honestly concerned for somebody other than herself? Earlier he had thought this impossible. Then he remembered what the Lord accomplished with Saul of Tarsus. Yes, anyone could change, with God's help.

"You seem troubled," she continued, her tone smooth and inviting. "Is there anything I can do to help?"

How was he to answer this? He didn't need or want Charity's help. But he didn't want to cast her off if she was sincere. And she certainly seemed so to him.

"Well," he began, "there is something—" He wanted the problem solved and thought Charity might be the answer. After all, she was Faith's sister. Sisters usually shared a special bond, even if personalities conflicted.

"What, Evan? Anything! Just say."

He hesitated. "Well, perhaps you *can* help. Faith won't speak to me. I think I've upset her, and I can certainly understand why."

Charity stood up, appearing to listen intently.

"I'd sure appreciate it if you would speak to her and tell her I'm sorry. Explain I was only doing what I thought best for her and that I would never have let her drown in the river. Never!"

"Of course. I'll do that for you, Evan. It's no problem at all. I'll explain everything, and I'm sure things will work out."

A sense of relief settled into Evan. He felt his shoulders drop as though the tension had left his muscles. This was good. Charity would do something considerate for a change, and Faith would know how he felt. They would talk, woman to woman, and he would finally be understood.

Charity handed the plate back to Evan. "Please eat. It would make me feel so much better." She paused. "And I'll speak with Faith real soon—don't you worry."

Nodding, Evan picked up the spoon and continued eating, while Charity headed back to the house and disappeared inside.

Evan glanced over at Faith then and noticed that both women were rolling their skins into tight bundles and tucking them into hide sacks. Sits Alone gathered her tanning tools in a *parfleche,* and she and Faith walked up the incline and also disappeared inside the house.

Making his way up the grassy knoll, Evan saw Mr. Lambert stretched out on the porch resting, his hat pulled down over his eyes. He was using a rolled-up gray army blanket for his pillow.

Mr. Lambert had taken the losses worse than the rest. His grief had settled in his eyes where a glow had once resided. Only after his daughters were brought back safely was the man able to function again. Even then, he moved about slowly and never smiled, keeping mostly to himself. Evan knew it would take a long time for the older man to heal. Each member of the Lambert family dealt with grief in his or her own way.

He thought about Virgil, with his youthfulness and resilience. He seemed to handle the deaths better than the others. Evan had caught sight of him many times, though, kneeling by the graves, talking softly as if in prayer. And Virgil worked constantly, probably to keep from thinking about it, Evan surmised.

And Faith. She had cried many times in the night, when she probably thought no one could hear. But Evan could hear when the night was still. And his heart ached to hear her soft moans and the agony lacing each muffled sound. He knew she missed her family. Then the sun would rise, and Faith would rise with it, holding tears at bay as they all became active again.

Charity was another story. She seemed all right, though she watched everyone's *every* move. He thought her anger toward Sits Alone was her way of expressing grief. Some people did that, became angry and tried to place blame somewhere, as though it might make them feel better. He shook himself from his thoughts and squared his shoulders.

He stepped carefully around the sleeping man and then inside to return the plate to Charity.

"You sure were hungry!" she exclaimed, with a gleam in her eyes

"Thank you, Charity," he replied. "I'm going outside and get some rest now."

Before he left, he glanced at Faith where she stood near the hearth. What a lovely sight. She never rounded her shoulders but stood straight and tall, her chin slightly lifted. She possessed a gracefulness he had never seen, from walking to the way she moved her hands. And she seemed to light up the entire room, even with the single oil lamp burning on the table, casting shadows about. With a picture of her firmly implanted in his mind, Evan left and sought a place to sleep near his horse—a place where he could dream of being married to the lovely Faith.

≈

Fort Garland was located twenty miles east of the Rio Grande between Ute Creek and Sangre de Cristo Creek. Some years earlier it had been Fort Massachusetts.

Uniformed infantry guards greeted Virgil Lambert at the gate and permitted him to enter only after a brief talk. Accompanying Virgil was one of the guards, Officer Warren, who wore a dark blue wool sack coat and light blue trousers. Stretched around his middle was a black leather waist belt, worn over the coat. On his head rested a dark blue wool forage cap with a looped horn ornamenting its face.

As the two men walked through the fort, Virgil studied its layout. He was curious to note the adobe brick buildings that enclosed a parade ground with the structures running along either side. He guessed these must house officers and scouts. Passing one of the buildings he caught the scent of sourdough bread in the midday air, stirring his hunger.

His attention was at once drawn to the activity in the middle of the parade grounds near the flagpole. Officer Warren answered his question before he could ask it.

"Paymaster just returned from Fort Union in New Mexico, and they're handing out payroll."

Virgil surveyed the soldiers marching in formation to the table as the paymaster called out name and rank.

"Most men walk away without a single shiner," Warren continued. "You see that man there and that woman?" he asked, pointing a bent finger that must have been broken at one time.

"Yes, I see them," Virgil said, nodding. He felt out of place at the fort yet transfixed by the sight of the soldiers, dressed in their uniforms.

"The man is the sutler; the woman is the head laundress." He cleared his throat and coughed. "We got to pay our debts first. Everything costs around here."

Virgil's attention was focused on the dark-haired woman, marking in her record book, as she took money owed her for doing each soldier's laundry.

"Who is she again?" he asked.

"Our laundress," Warren repeated. "We've got four." He coughed again, this time holding his belly as he did so, and then spit.

"Tough job, I guess," Virgil said. He knew she was the head laundress; he had wanted her name.

"Probably the hardest job around these parts. Most of them are married women, with children." He paused as though he were going to cough again but didn't. "That one there, that you keep eyeballin', just lost her husband this past winter to scurvy. The sutler was out of vinegar and pickles, which would have helped. Lost four good men to the sickness. That woman's husband was the last to go. No more losses since."

"Sorry to hear that," Virgil said, a burning sensation rising in his throat. He understood that kind of loss all too well. Maybe her name didn't really matter. He felt as if he already knew her in some strange way.

Another name was called, and the line moved forward.

"Poor woman's had a barrage of suitors tossed on her all at once. But she isn't havin' any of it."

"She needs time," Virgil said, his compassion suddenly turning to anger. How could those men think only of themselves when the woman was in mourning?

"Pretty little gal, wouldn't you say?"

"Yes, she is," Virgil agreed.

Just then he noticed her glancing his way, and he raised his hat in her direction. She smiled and returned to collecting the laundry money. The line of soldiers was dwindling.

"Well, if that don't beat all!" Warren exclaimed. "That's the first time I've seen her smile since she took to being a widow." He looked at Virgil. "And she smiled at you!" He chuckled. "If you aren't the lucky coot! How old are you, anyway?"

"What's her name?" Virgil asked, ignoring the man's question.

Virgil was struck, even at that distance, by the young woman's eyes. She possessed a natural beauty about her, with her dark brown, shoulder-length hair twisted loosely into a widow's knot at the nape of her neck and a few strands falling from it.

"Doris," Warren answered. "Doris Randall, young widow of the late Private Ralph V. Randall."

Virgil spoke her name softly under his breath, enjoying the feel of it on his lips. He continued to watch her, captivated, before he recalled why he had come to the fort.

"I need to talk with someone about the Indian attack," young Lambert said, suddenly facing the guard.

"That'd be Lieutenant Conline," said Warren. "Follow me. I believe he's in the sutler's store."

The two men passed the company quarters, two long buildings one story high, with thick adobe walls topped with mud roofs, until they reached a building set off by itself. Inside, the scent of pine and earth hung thick in the air, from the pine board ceiling covered with earth and the plank board flooring. Mud-plastered walls had been white-washed with lime and bore buffalo, bear, and deer skins.

Also on the walls were Colt and Remington pistols lined up with Springfield and Sharps rifles and bayonets and sabers. Saddles rested in the corners of the store, and a variety of Mexican spurs, stationery, cosmetics, and ammunition sat on the top of a counter.

The sutler was standing behind the counter with another uniformed officer.

"Sir, I've brought Virgil Lambert in," Warren said, introducing the men. "He says he has a story he needs to tell."

After sending Warren back to resume his post, the lieutenant turned his attention to Virgil, who told the story of the attack on his family.

"I'm sorry to hear about your losses, Son," the gray-haired lieutenant told him. "We're having more and more problems with the Indians. Wagon trains, settlers, even forts and trains are being attacked maliciously and apparently without cause."

"I believe the Indians who attacked my family felt they had cause."

"No, Son, you can't justify that kind of brutality."

"I realize what you're saying, but they thought my brother James and the former owner, Mr. Rucker, were one and the same." He gave the facts as he understood them.

The lieutenant studied the young man. "I'll dispatch a unit up your way," he said finally. "We need to send word of the attack to Washington and to surrounding towns and forts." He paused. "You sure it was Cheyenne that attacked and killed your family?"

"Yes."

Lt. Conline looked out the window briefly and then back at Virgil. "Why don't you bed down in the infantry barracks for the night? You can head out in the morning with a company of my best men."

Virgil nodded. "Thank you very much, Sir. I'm much obliged." He did feel tired and hungry.

"You're free to walk about the grounds if you like and take your dinner in the mess hall. There's enough sun left to look around."

"Thank you again, Sir."

Virgil stepped outside into the air. He knew what he wanted to do, and it had nothing to do with eating. He had brought enough food to sustain him for days. No, he wanted to find the young woman named Doris. She had captured his attention with the first glance. Just how he would make her acquaintance he didn't know. And once he found her, if she showed no interest, he would back off. But Virgil was certain her smile had meant something, and he intended to discover what.

nine

A sound too distant for the average ear caused Faith's heart to quicken. She rose from her squatting position where she had been fleshing new rabbit skins. Sits Alone had completed the process on the skins started two days earlier.

"What is it, my sister?" asked Sits Alone.

Faith tossed her head back, arching her neck and inhaling deeply.

"Riders," Faith answered. "Several." She straightened her neck and breathed deeply again.

"Who? Enemy?"

When Faith didn't answer immediately, Sits Alone called out loudly to Evan. He came running, rifle in hand. Mr. Lambert was off on a hunt again, this time with Barnabas, hoping to bring in larger game for the winter. They had promised not to be gone long or go too far away.

"What's wrong?" Evan asked, catching his breath.

Just then Charity ran up to them, her eyes wide, her breath coming in short gasps. She had been napping on the buffalo hide in front of the hearth.

"Faith hears something," Sits Alone explained. "And I think she smells something too."

Evan scanned the area. "I'd better go have a closer look."

"*Noxa'e,*" said Faith.

"What did she say?" asked Charity, surprise in her voice.

"She said for us to wait." Sits Alone smiled. Her student was learning fast.

The others were unaware that Faith was studying the Cheyenne tongue and, with her good memory, had made rapid progress.

"It's okay," Faith assured them. "It's not the enemy. I smell my brother and his horse. But I smell other horses too."

"Indians!" Charity nearly screamed the word, as if the word itself had taken on flesh and blood.

"No, not Indians," Faith said, with irritation in her voice. She regretted her sister's behavior toward Sits Alone.

Evan leaned the rifle against the porch railing. "It's Virgil," he told the women. "And it looks as if he's brought some help from the fort."

The riders stopped some distance from the homestead and exchanged a few words. Then the group divided, the soldiers going one way, while Virgil and another rider continued to the house.

"There's someone with Virgil," Charity said sharply. "Who is that?"

"I smell—" Faith began, but she couldn't finish.

"A woman?" The answer seemed more a question than a statement as Evan uttered it.

Faith heard him step off the porch, the dirt crunching beneath his boots.

She had known it was a woman before Evan said anything. An unease settled over her like a scratchy blanket.

Silence followed except for the sound of the approaching horses. Within minutes the horses came to a stop. Virgil was the first to dismount.

"Faith, Charity, Sits Alone, Evan," he began, "I want you to meet Doris."

"Why did you bring *her* here?" Charity blurted out.

"*Hekotooestse!* Be quiet!" Sits Alone said sharply to the girl.

"Glad to make your acquaintance," Faith said, hoping to hear the newcomer's voice.

"I–I'm very happy to meet all of you," Doris said, with some hesitation. "Virgil's told me—so much about you." She glanced about as if looking for someone else.

"Oh, my Pa and Barnabas are off hunting. We're trying to get ready for the winter months ahead of time since we don't have enough season to plant crops to carry us through."

"I'm—anxious to meet them, Virgil," she said, apparently nervous. "And I certainly feel privileged to meet the others."

Her voice sounded sweet and genuine to Faith, and she wondered if she could sing. She also couldn't help wondering what Doris was doing there and why Virgil had brought her.

❧

Later that evening at dinner, after Mr. Lambert and Barnabas had returned, they were all sitting around the pine table. As usual Barnabas asked the most questions.

"How long were you the head laundress, Ma'am, at the fort?"

"About seven months," Doris answered. "It doesn't take long to move up. People don't stay long, it seems. The threat of attack and the workload make many people wish to be back at their homes."

Faith ran her fingers over the rough fibers of pine in the table. She started to ask a question, but Virgil spoke first.

"Sickness makes people want to leave too," he added. "Unfortunately, Doris lost her husband to scurvy. Others died as well."

Everyone grew silent. To Faith the quiet seemed to last forever.

"I am sorry for you," Sits Alone said, breaking the silence.

Faith understood. The poor young woman had been alone, after losing her husband, and probably longed to be home again.

"Thank you," Doris answered but said nothing more.

Faith noticed that Charity hadn't said anything since her earlier outburst, and she wondered what the problem was. But she dismissed the thought and focused her attention on the visitor.

"Do you have family?" Faith asked, hoping she wasn't causing Doris more sadness. Maybe the young widow was on her way to reach her kinfolk.

"No, not much," she told Faith. "Well, I do have an uncle, but I don't know where he is." She was quiet again. "I lost my folks to smallpox when I was about ten years old. I was raised by an old woman who took me in. But she's gone now too."

Not sure what to say, Faith echoed Sits Alone's earlier words: "I'm sorry."

She thought she understood now. Her brother had brought the young widow here because she had no one and no place to go. She was as alone in the world as Faith felt at times. Her heart was full as she sensed her brother's compassion. Some questions remained, but Faith believed they would be answered in time. Right now she resolved to make Doris feel at home.

❧

"Faith, I need to speak with you."

Charity's voice brought Faith back to the present. It was the next morning, and she had been reviewing in her mind the process of tanning hides, trying to commit it to memory.

At her sister's words she pulled the quilt tighter around her shoulders to ward off the morning chill and continued rocking in her mother's slat-backed chair.

"Did you hear me?"

"Yes," Faith answered. "Yes, I did. What is it?"

"I really need to talk with you—it's important."

Charity sat down in front of Faith on the bare floor. Having just bathed, she smelled fresh, like lye soap. Still, Faith couldn't help but feel uneasy. She wouldn't feel this way with anyone else in the family. But Charity had never said she needed to talk with her.

"What is it?"

Charity hesitated, swallowing hard. "I–I have something to say." Her voice sounded constricted, as if she were about to cry.

Faith found herself wanting to comfort her sister, mend whatever was upsetting her. "If there's something wrong, Charity, we can fix it. Are you in some kind of trouble?" Her compassion always overcame her discomfort.

"Oh, no, no trouble. But something has been bothering me. Two things, really."

"Okay, what's the first?" Faith asked, leaning closer to her sister.

"It's just that I've been feeling ashamed lately."

"Ashamed? Of what?"

"Well—because I haven't been the kind of sister to you I should have been."

Silence fell over the room.

Finally, Faith found her voice. "I think you've been a wonderful sister to me, Charity. We've had our difficult moments, but—"

"No, you must let me finish."

"Of course."

Charity shifted her position on the floor. "You see, the loss of our family has made me understand the importance of family."

Faith fought to hold back her emotions. She felt like laughing and crying for joy at the same time. But she knew she must let Charity say what was on her heart.

"You, Pa, and Virgil are all I've got left. And I know I've been horrible to you, Faith. I haven't behaved one bit as a sister should. And—and I'm so sorry for that—for my horrid actions—for the way I've treated you. And I'm sorry I slapped you that night back in Nebraska. That should never have happened, and it never will again."

She drew in a deep breath. "All I ask is your forgiveness. And if you choose not to forgive me, well, that's my own fault, and I'll certainly—"

"Of course I forgive you, Charity!" Faith could no longer hold back. "I've already forgiven you—before you even thought about asking." *Oh, Father! This is an answer to my prayers!*

Charity continued, as though Faith hadn't said anything. "I'll certainly understand." She paused. "Do you remember how I drew closer to the Lord, after we'd been captured by those Indians?"

"Yes."

"Well, I realize now that I should draw close to the Lord, no matter what the circumstances."

Pressing both hands against the sides of her flushed cheeks, Faith said, "Yes, Charity, that's right! We should walk with the Lord in the valleys as well as over the mountains." *How wonderful,* she thought.

Faith felt Charity take her fingers from her face. Her touch was cold.

"I want you to know I've changed, Faith. Do you understand that?"

"Of course, my dear sister. I understand." She felt excited inside. "I'm so happy to have my precious sister back!"

"And I'm so happy to be back!" Charity's grip tightened around Faith's fingers.

At this, Faith leaned forward more, to embrace her sister, but Charity resisted.

Faith pulled back at once, confused. "What is it? What's wrong, Charity?"

Charity drew in her breath again. Faith waited, not wishing to upset the balance of such a fragile conversation.

"Well, Faith, remember I told you there were two things I had to share with you?"

"Oh, yes, I almost forgot." Her wrinkled forehead relaxed.

The rustling of skirts and the sound of boots scraping against the hardwood floor told Faith her sister had moved in very close. She felt her warm breath on her face.

"What is it, Charity? You can tell me anything. We're sisters, remember?" Faith's heart pounded.

"Yes, we're sisters. That's why I must tell you this—because I love you so dearly."

"Whatever it is, we'll get through it together. I promise you that." Faith tried to steady her pounding heart. "I will be with you every step of the way. Don't you worry." *That's what family is about,* Faith told herself, *being there for one another no matter what the circumstances.*

A slight smile formed on Charity's lips, though Faith couldn't see it. "No, Faith, I will be with *you* every step of the way," she said, her voice unchanged from the earlier conversation.

Faith felt the tightness between her shoulder blades, and her temples were throbbing.

"Please, Charity, tell me what's going on! I must know!"

"Shh, keep your voice down," the younger girl whispered. "Nobody else must hear this. It's personal, Faith, and I don't think it's anybody else's business."

"Sorry. Please continue."

"Okay, but try not to get too upset."

"I'll try." *Please get to the point.*

"I've been hearing bits of conversation—about you."

"About me?"

"Yes, and I think you have a right to know what's being said." She hesitated. "Also, I was asked to speak with you, by a person you care dearly about."

"Who?"

"Evan."

Faith managed to keep her hands clasped tightly in her lap, her fingers clenched, instead of raising them to her face, as was her habit. She felt the tension in her knuckles.

"What did he ask you to say to me?" Faith asked sharply.

The silence that followed only increased her agitation.

"Tell me," she demanded.

"You must calm down first," Charity told her.

Calm down! How can I calm down? "Okay, I'm calm," she said, knowing she was far from calm. "Now will you tell me?"

Charity looked around the room and out on the porch, to be certain no one else was listening.

"Evan said he would tell you himself, but he couldn't bear upsetting you. So he asked me if I would help him out."

"Help him with what?"

"You shouldn't blame him for this, Faith. Do you understand?"

"I don't even know what I might be blaming him for," she said, with growing irritation.

"You know what happened at the river—when you almost drowned?"

"Yes."

The words were coming more easily now. "Before that happened, I overheard Evan talking with Barnabas—in quiet tones, mind you."

"Go on."

"He likes you, Faith. Did you know that?"

"You mean Evan?"

"Yes."

"I had that impression."

Charity paused, letting her breath out slowly. "And I guess you've taken a liking to him too. Right?"

Faith shifted her position in the rocker, causing it to creak. "Yes," she finally said.

She grew more uncomfortable with the direction of the conversation as each minute passed and more words were spoken. Yet she knew she must hear them.

She was confused about one thing though. Never would she have imagined that Evan would confide in Charity. The two never got along. What had changed?

Faith recalled her sister's earlier pleas for forgiveness. Perhaps her heart had yielded to a new understanding. Faith had to believe the change was sincere. Maybe she had spoken with Evan also, and now he felt comfortable confiding in her.

After all, she spoke with the greatest sincerity. Faith could not detect anything false in her tone. Besides, why would she play games with something like this? Even Charity had more heart. No, the change had to be real, Faith surmised, probably brought about by the trauma they had just survived. It took that for some people, and it must have for Charity. If anything good could come from their losses, Faith decided, this was it—a new beginning for her sister as a Christian.

"I guess Evan thought it might work out," her sister said suddenly, interrupting Faith's thoughts.

Charity stood up and began pacing sharply back and forth across the hardwood floor. Her steps were muffled when she reached the buffalo hide in front of the hearth but were clear again when her shoes landed on the bare plank floor.

"I overheard him talking with Barnabas," she went on, "about you."

"About me?"

"Yes. I didn't mean to listen in on their private conversation, but when I heard your name, I couldn't help myself."

"What did you hear?" Faith asked, not so much concerned for Charity's feelings of guilt as wanting to know what was said.

"I heard Evan tell Barnabas that he wanted you to be his wife."

Faith's breath caught in her throat. She wiped away the moisture that had gathered on her forehead.

"Wife?"

Faith remembered their earlier conversation, when she had to turn Evan down because she felt inadequate. She was unaware anyone else knew about his request.

"Yes, Faith, but don't get too excited."

"Please go on."

"Well, he asked Barnabas what he thought. I guess Barnabas wasn't too keen on the idea, and he gave Evan some advice that I wasn't too keen on."

Clenching her hands tightly in her lap, Faith waited.

"Barnabas told him you could never be of good use to him or any other man for that matter. Of course, Evan balked at the notion at first."

Yes, thought Faith, *that sounds just like Evan.* Yet it hurt her that Barnabas would feel that way about her. That didn't sound like the Barnabas she knew. Perhaps she'd been wrong.

"Barnabas told Evan to watch to see if you could complete a task—do something without the help of others. That would answer his question about a life-long commitment with you."

Tears threatened as Faith recalled the river incident. As if she knew what Faith was thinking, Charity went on.

"After you nearly drowned in the river, I overheard them talking again. Barnabas told Evan he had his answer. You simply did not have it in you to know independence. And you would always be dependent on somebody for all of your needs."

Now the tears flowed freely. A cold sensation gripped Faith's heart. She felt pressure as Charity wiped a tear from her flushed cheek. She thought it odd that she felt no compassion in her sister's touch. But perhaps her emotions were in the way of any other feeling at present. Faith fought to stop the tears.

"That's why Evan didn't help you, Faith. I guess he wanted to be sure you *couldn't* do what you'd set out to do."

"So he would have let me drown?" *Of course he wouldn't. How can I think that?* she chided herself. She felt confused.

"I don't really know, Faith. I certainly wouldn't think so, would you?"

"No, but he didn't offer to help, and h–he did try to prevent Virgil from helping me. I–I just don't understand this. Or–or maybe I do."

In some way Faith thought she did understand, but that didn't ease the pain with it.

"I have something else to say, Faith, and I hope you won't be too upset."

"Go ahead." What did it matter now? Charity could say anything, and it wouldn't increase the terrible ache in her heart.

"Evan wants to court me," Charity replied, nearly whispering.

The pain shot through her whole being.

ten

For a long moment Faith could find no words. She knew she should be happy for her sister and for Evan. After all, hadn't she balked at Evan's idea of marriage? Didn't she mean it? So whose fault was all this? Faith unclenched her fingers. Evan certainly had a right to find someone else, though she never dreamed it would be Charity.

"I want you both to be happy," she managed to say at last.

It was true. Faith did wish them happiness. But she couldn't sort through her feelings now. She had to get through this one moment and then work on the next. *Lord, help me!*

"Do you really?" Charity asked, her voice rising.

"Of course."

A laugh escaped Charity's lips. "I knew you'd under-stand!" she said with a sudden gleeful tone in her voice.

How strange, Faith thought, *that she can change her emotions so quickly. When we first started talking, she seemed nearly in tears, but now—* No, she must stop this.

"I'd like to be alone," she told her sister.

"Oh. Of course." Charity's tone was steady and smooth again. "How rude of me. I know you need some time, Faith, and I'll certainly give it to you."

"Thank you." Faith longed to release the tears welling up inside.

"If you need anything," Charity continued, "anything

141

at all, you call me. Okay?"

"How sweet, Charity. Thanks."

"That's what sisters are for. I've changed, and I want you to know that."

I know, and apparently Evan does too. Once again Faith tried to force the negative thoughts from her mind. Her sister had been born again in Christ, and she should be rejoicing!

She reached out both arms, and Charity leaned into them. Faith found her sister's cheek and kissed it gently. "I love you," she whispered, "and I will be praying for you and your new beginning."

Charity returned the kiss and embrace and then dashed out of the house, singing as she went.

Rising from the rocker, Faith stepped over to the window. There she let the morning sunshine warm her skin. Tears ran down her cheeks and onto the lace collar of her dress.

She pulled in a shuddering breath. "Father, I thank You so much for coming into my sister's heart, for bringing her to salvation, and for bringing her back to us! She is part of our family, and I am so thankful."

She remembered the Scripture verse about leading the blind and making their paths straight and thought it must surely be meant for Charity and not necessarily for her. Charity had been brought by a way she knew not, and her crooked path was now straight. She should be happy—and thankful!

"Forgive me for my feelings, though, Father, for I can't seem to control them. I am truly thankful. But now I ask for guidance as I move forward. Help me learn, help me become independent, as I must surely be alone in this life!"

Quickly she brushed the tears from her cheeks. "I'll do the best I can, though, with Your help. You've given me a wonderful teacher in Sits Alone. And I'm grateful for that. But I ask now for understanding of all You're trying to teach me. And I ask, if it's Your will, that You'll ease this pain in my heart. I'm human, and sometimes I don't understand. I ask for wisdom, just as Solomon did. I promise I won't abuse it. Thank You, Father. Amen."

As Faith completed her prayer, she heard laughter outside. It was Doris and Virgil. They talked in low whispers and then laughed out loud. Faith also heard the happiness in her brother's voice. He had found someone. She realized the Lord had been working in their lives too, his and Doris's. Virgil would not know the loneliness she herself knew. And Doris had found a home and a companion as well. Then there were Evan and Charity—

Faith closed her eyes, feeling the heat behind her lids. Not caring, she brought both hands up and pressed them against each side of her face. The action brought her some measure of comfort. After a moment she withdrew her hands and stepped to the rocker. She would sit and wait for Sits Alone to come for her to begin the morning routine.

❧

Virgil watched Faith for the next two months, all the way into August, and was astounded at her accomplishments. She had far exceeded what he thought she might be capable of. She walked to the river on her own, filling water pails or washing dishes. She tanned hides and sewed shirts and leggings, even a pair of moccasins. And her beadwork was better than Sits Alone's, and the Indian woman could see!

Faith guided each bead or quill into place with the tips of her sensitive fingers. And no discerning eye would ever know the work was done by a blind woman, unless told.

Life had gone on, and they had survived the past. The army found no sign of renegade Indians in the area and had drawn the conclusion that it was a single act of revenge. Unfortunately, the innocent had paid. They never learned who the killer was and probably never would. Virgil had to accept that and move forward. The sad part was that the army knew the identity of the white man who had killed an innocent warrior. They knew who he was and where he was. Yet they did nothing about it.

Barnabas and Sits Alone had journeyed to their own home once, to make sure everything was all right. They had returned with two pack mules to continue assisting the Lamberts. Evan had stayed too. He slept under the stars every night, working diligently and keeping mostly to himself.

Virgil discovered he loved Doris and professed this. The two planned to wed. It occurred to him, as he was building his own relationship with Doris, that he now understood Evan's position. He knew he couldn't keep Faith and Evan apart any longer. Love was more powerful than he had ever imagined. Admittedly he was baffled though.

Faith showed no interest in Evan. Virgil had been afraid of this, that Evan had traveled all this way, longing to win Faith's heart, but failed. He shook himself and continued the task of helping the women cut deer meat into thin strips. They would hang the slices to dry so the family would have a plentiful supply of jerked meat once winter arrived.

&

"Evan, I brought you dinner," Charity said, offering him the plate.

The sun still hovered above the edge of the fiery skyline. Evan shifted his sitting position and felt a pine needle prick him in the calf. He pulled it out and leaned back against the rough outer edge of the pine tree, his legs in front of him, crossed at the ankles.

"Thank you," he said, taking the plate of food from the girl.

"What's wrong, Evan? You never seem happy anymore."

"I'm fine," he told her brusquely.

"But you've been like this ever since I confided in you, after doing what you asked of me."

Charity dropped to the ground in front of him, drawing her knees up toward her chin and wrapping her pale green skirt around her legs. She smoothed the dark green lace on the bodice of her dress with one hand.

"I feel as if you're angry at me, for the way things turned out—for Faith's decision."

Forcing a smile, Evan studied Charity's large brown eyes. It wasn't her fault, he knew. But he had never expected this turn of events. Maybe he had simply hoped for too much.

"I'm sorry if I've made you feel that way," he told her.

"Oh, but you have a right to be sad. After all, you do love my sister. I understand that completely!"

"Yes, I do love Faith. I will always love Faith." *That's the problem!* he told himself.

Evan thought he saw something flicker in Charity's

eyes, but it was gone before he could guess what might have caused it.

"Is there anything I can do?" Charity asked, smiling.

"You've done enough," Evan said. "You tried to talk with Faith, as I asked. And you've made sure I've been well fed." He smiled. Charity really had been a jewel, and he wanted her to know that. Her change of heart certainly seemed genuine. "I'm very grateful to you."

Charity wrapped her arms around her legs and rested her chin on her knees. "It's the least I can do," she said softly.

Evan chewed the sweet corn pone slowly, thoughtfully.

"Don't you want my sister to be happy?"

Evan coughed hard to dislodge the bite that had suddenly gone down the wrong way. He managed to clear his windpipe.

Is that how everyone sees me? he asked himself.

"Of course I want Faith to be happy," he said, finally giving voice to his thoughts. "That's the most important thing." *Surely this family knows that!*

"Oh," Charity said, "it just doesn't seem that way."

"Why not?" he asked abruptly. He didn't want to discuss this anymore; the girl's frankness made him uncomfortable.

Charity smiled, exposing perfect white teeth. "You just mope around, as though you're not moving on with your life."

Faith is my life! Evan knew he couldn't say that out loud.

Maybe Charity was right. If she believed her own words, then the other family members must conclude the same. This wasn't good.

"I guess I never expected Faith to choose a life alone. I

'thought she must have some feelings for me.'

"Oh, but she does—or did. I mean—"

Evan raised his hand to halt further words. He hadn't completed his explanation.

"Besides, you said it was because of her dependence on everyone—yet look at her!"

By that time Charity was gripping her legs tightly. "Yes, but—"

Evan silenced her with another wave of his hand. "She can cook and clean and sew—better than most." He paused and swallowed. "Her ability to learn is incredible, and her ability to overcome. So—if she was worried about being a helpmate to me, she can worry no more."

"Well, that wasn't all."

"What do you mean? I thought you said—"

"It's true what I said, but there was more. I just wasn't sure you'd be able to handle the rest of the truth."

"The rest of the truth?"

"Yes."

"Go on."

"You see, Faith had already decided she was going to learn to be independent. That's very important to her, as you know."

Evan nodded, not sure of what was to come.

"She also believes that if she were to marry, she would lose that independence, that her husband would still feel as if he must do everything for her. She doesn't want that, Evan. She would rather be alone than lose her independence."

Evan set the tin plate on the ground beside him. "I wouldn't do that."

"But she doesn't know that. She wants her independence so badly that if life alone is what brings it, she would rather be alone." Charity released a suppressed breath. "Does this make sense?"

"I guess so."

"But that's not all, Evan. I guess it's up to me to tell you the whole truth, no matter how much it hurts."

Evan felt the muscles in his legs twitch and a hot sensation fill his stomach.

"You see, Evan, she never really has had a love for you, not the kind you have for her."

"What does that mean?"

"Well, do you remember in Nebraska when you asked Faith to dance?"

"Of course I remember." *How could I ever forget such a wonderful time?*

"Well, she's hated dancing ever since. And she's lost her love of flowers."

"Flowers?"

"I think it was because you put that daisy in her hair. The whole experience was very bad for her."

But wasn't it you who made it bad for her? he wanted to ask, but he chose not to. Maybe there was more to it. Maybe he had pushed her into an experience for which she wasn't ready. Suddenly he wondered if he had gone about everything the wrong way.

"She refuses to touch flowers, and she won't dance again. She told me so." Charity shifted on the ground and curled her legs, tucking them beneath her.

Evan didn't know how to respond. His heart ached for the misery he must have caused Faith. How could he have been so blind to the needs of the one he loved more than anything else on earth?

"I must apologize to her!"

"Oh, no!" the girl said at once. "That would only make matters worse! I know my sister. You must trust me on this."

Evan wasn't convinced, but he didn't tell Charity. She seemed concerned for her older sister, and he didn't want to upset her as well.

"Faith is aware of your unhappiness, and this makes her unhappy. That's why I'm telling you about this. I know how badly you want my sister's happiness."

"Yes—yes, I do. And I will do better. I wasn't aware of my own actions. My own thoughts, maybe, but not how I appeared to everyone else." *Perhaps apologizing would make things worse.*

"That's understandable." Charity played with a pine cone that had fallen to the ground. "I think there's somebody else out there for you, Evan. Just because my sister wants a solitary life doesn't mean you have to endure one."

Charity appeared to have changed in many ways, but in others she still lacked understanding. Evan had no desire to teach her. His heart was broken. But if agreeing might help her, then he would.

"I guess you're right. It's time for me to move on. Maybe there *is* somebody else out there for me." He knew, as the words left his mouth, that they weren't true. Faith was the only woman he would ever want. But Charity could never understand that until she found the right person for her.

For the first time Evan thought about leaving the Lamberts. He knew he couldn't change his demeanor. He could try, but his heart would eventually dictate his attitudes and his actions. Yes, if it meant Faith's happiness, he might have to move on.

eleven

Sits Alone entered the dwelling. Faith had cleared out the ashes in the hearth and was filling a second bucket when she heard the door open.

The scent of her friend brought an instant smile to Faith's face. She lifted the bucket, walked to the door, and set it down. "I've already swept the floor too," she said. "You can take a look if you want."

"I trust you did the job well," said Sits Alone. "No need for me to look."

Something in the Indian woman's voice sounded strained, distant almost.

"Are you all right?" she asked.

Faith heard Sits Alone walk softly in her moccasins to the buffalo skin rug. By the gentle rustle of fur and buckskin sliding together she knew the woman had settled onto the rug.

Something wasn't right. Faith was sure of it.

"*Nenaasestse.* Come here," Sits Alone said.

Faith wondered if she wanted her to tell her a Bible story. She had returned the Indian woman's teaching favors with stories. Sits Alone, unable to read, wanted to know the Bible, and Faith, with her excellent memory, had recited many of the stories and parables to her during the evening hours.

Together the two women had discussed verses in the

Psalms and Proverbs, encouraging one another, and Faith had told her friend about Moses, Noah, David, and other characters from the Bible. They had discussed Paul's journey and his wonderful letters and his steadfastness, even to his death. Usually, however, they had enjoyed these times after the last meal of the day, when the sun had gone down, not at this time of the day.

Something told Faith that Sits Alone didn't want another story. She crossed the room to where the woman sat and seated herself in the rocking chair.

"Okay. What's wrong?"

"There are some things I must ask you," the woman said.

"I'll answer the best I can."

"I saw Evan and your sister talking," she told Faith. "I've been watching them for some time."

Faith thought she understood Sits Alone's purpose in speaking with her.

"It's okay," she said, her voice quivering slightly. "Evan and Charity have been courting." She hesitated. "They love each other."

"No! They do not!"

Faith's fingers tightened on the arms of the rocking chair. Sits Alone apparently didn't understand what had transpired.

"It's okay," she said again. "Charity and I had a long talk about two months ago. I'm perfectly content with this, and I wish them the happiness they deserve."

"Even when your sister takes happiness away from you!"

"It's not her fault. You must try to understand—"

"No, it is you who must understand," Sits Alone said.

"What she has told you is not truth. They are not—'courting,' as you say."

"What? What do you mean?" Faith felt the muscles in her neck tighten.

"I must ask some things first," Sits Alone said.

Faith waited.

"Do you like the things Father has given us?"

"Of course I do," Faith answered. "The Lord has given us many wonderful things."

"Do you enjoy the flowers, trees, and plants?"

"Yes." Faith wanted to tell Sits Alone how she loved the smells of various plants, but she didn't. She also wanted to describe to Sits Alone the time Evan had slipped a daisy behind her ear and told her it was the same color as her hair.

"Have you ever danced before?"

"Yes." Faith felt confused by the strange questions. "Once—but only for a short time." Again she wanted to tell her friend about Evan, but she held back. She could feel her heart pounding.

"Do you hope to marry someday and have children?"

"No—I mean, yes. No," she stammered.

"Are you speaking the truth?"

"No." Faith covered her face with her hands, trying to press back the tears. She swallowed hard and drew in a deep breath, trembling with the effort. "Stop, please! I don't know what you're doing or why!" She dropped her hands into her lap.

"It is because I love you, my sister."

"I still don't understand." Faith's chest rose and fell with another shuddering breath as she tried to calm herself.

"Then let me tell you," Sits Alone said.

At that moment they heard horses riding up.

"They're back!" Faith exclaimed. "And I can smell that the hunt was successful."

Both women rose to their feet. Faith walked to the door and stepped outside, followed by Sits Alone.

"Papa!" she cried, relieved for the interruption.

Mr. Lambert bounded up the steps to Faith's side and wrapped one arm around her. "I sure am glad to see you, little girl, but I'm mighty grubby. We've already gutted these critters, since it was gettin' dark."

He looked at Sits Alone, who stood calmly beside her husband.

"You can wait 'til morning, if ya want, before ya start on these skins."

Just then they heard another horse approaching. Faith smelled the familiar yet subtle aroma of cinnamon, as well as the rich, musty scent of Evan's mustang. Her brows raised, as if to ask a question, but she said nothing. Whatever he was doing was his business. She didn't need to know. It wasn't her habit to ask questions anyway.

She heard a hand slapping buckskin and guessed Evan had given Barnabas a solid pat on his back after he dismounted.

"Good job, Man! You brought back even more than last time. You'll all stay well fed for the winter months and into spring at this rate!"

"Where're you heading?" Barnabas asked, ignoring the compliment.

Faith was glad Barnabas wasn't shy about asking questions. She listened closely for Evan's answer.

He paused, and before he could say anything Virgil and Doris wandered up to the group gathered in front of the porch. They greeted the two men and bantered back and forth several minutes about the events of the hunt.

Faith was growing restless. Evan was apparently leaving. But for how long? Where? And why?

Her legs grew weak, and she thought her knees would buckle under her. Suddenly she sensed that Sits Alone had moved to her right side. The warmth that pressed against her gave her strength.

Something was wrong, though. It charged the air about them like a bolt of lightning striking close by. Faith sensed it, and she knew everyone else did too.

And then she heard another horse approaching the group.

"You're back!" Charity exclaimed.

"What're you doin', Girl?" Mr. Lambert asked, with a puzzled tone in his voice. "And what do you have Nat geared up for?"

Charity laughed. "I'm going with Evan, of course," she told her father.

Everything was as clear to Faith now as it had been in the months that followed the intimate talk she'd had with Charity. And so were the emotions that had piqued during that conversation. She had to fight to control them now as she had then. Faith hoped Sits Alone would understand now and let things be.

The Indian woman walked over to Charity, where she stood holding Nat's reins to keep him from grazing on the bunch grass. Sits Alone reached up to stroke the horse's muzzle. The animal snorted and accepted the affection.

"Where is it you are going?" she asked Charity.

"I said I'm going with Evan."

Faith could hear the tension in her voice. The two women had never gotten along.

"I thought we had planned to head out without saying our good-byes," the girl said, apparently directing this to Evan. "We both agreed that would be best. Didn't we?"

Instead of answering Charity, Evan spoke to the others.

"I'm heading to Fort Garland to see if there's work. I'm very grateful for all you've done for me and wouldn't have traded these months with you for anything. But I think it's time I move on."

Faith could scarcely catch her breath. Hadn't something told her this day would come? She had subconsciously pushed the thought away, afraid to face the harsh reality.

Evan and Charity had developed a relationship and were now ready to venture off into a new life. She guessed it was probably Charity's idea to leave the home. Her sister was always anxious, always wanting to experience all the world had to offer—good or bad. Now her dreams would come true—with the man she loved beside her. And—with the man Faith loved.

Evan's voice cut into her thoughts. "Charity will accompany me to the fort, to see if that job is still open, the one Doris had."

"And will you be married when you get there?" Sits Alone asked.

"Of course they will," Faith said, trying in vain to sound happy. She wasn't prepared for what came next.

"Of course we will not!" Evan's voice had changed.

He sounded angry—and something else Faith couldn't discern.

"We don't need to get into all this," Charity said hurriedly. "Misunderstandings are not good to have right before a good-bye. We really should be heading out." She pulled up the horse's reins and started to mount.

"You should head out in the morning," Barnabas said firmly, "when it's daylight."

"It was my idea that we go ahead and leave now. I thought we could make camp after we'd traveled awhile, and then we'd be that much closer by morning."

"It's a good journey to the fort, Woman," Barnabas pointed out. "Best to wait for morning."

"When will you be married?" Sits Alone asked a second time. She was standing beside Faith again.

The scent of cinnamon grew stronger. Faith guessed Evan must be chewing harder. He was apparently distressed. And she was confused.

"Married?" Evan repeated the word. It hung in the air like a troublesome mosquito.

Faith heard her sister shuffling her feet in the dirt. Suddenly she understood. She had been deceived. They had all been deceived. Her heartbeat seemed to slow to a sluggish rhythm as the awful truth hit.

"What is this all about, Charity?" Evan demanded, anger rising in his voice.

"Evan, we must leave! Now!"

"Charity said you had been *courting*," Sits Alone said slowly, evenly. "We did not want to miss the ceremony."

Faith could hear the low growl in his throat before he uttered a word.

"What's this all about, Charity?" he asked the girl again.

"I did it for your own good," she told him, her voice

quivering. "Faith could never be a wife, much less a mother! And I can be all those things. You would see."

"No, never! You can never be any of those things," Evan said sharply, "without the right kind of heart—the kind of heart Faith has."

Faith struggled to keep her own temper under control. What good would it do? They had lost so much already. Most important, Faith realized, they had lost Charity. Her heart had not changed. Faith's lower lip trembled, and her eyes grew hot with tears.

Suddenly, Faith knew Evan was standing in front of her. She had the same wonderful feeling she'd had back in Nebraska at the dance. She sensed the comfort of his presence around her.

"Faith," he whispered, "I must ask you something, but it will mean something else."

She didn't know what to say. She was caught between two emotions: her sister had deceived them, and Evan was a free man. One emotion hurt deeply, while the other made her feel wonderful.

It didn't take her sister long to speak, however.

"What are you doing, Evan? I thought we were leaving here and getting as far away from these people as possible!"

Faith heard shuffling of feet and a grunt.

"Take your hands off me, you big oaf!"

"You need to be quiet and still, little lady, while these poor folks try to undo the damage you've caused."

She heard a whimper from her sister and the horse snorting and stomping about, as if to warn Charity as well. Then all was quiet.

Faith's heart was pounding in her chest. She drew

in deep breaths to steady herself. Again Evan spoke to her.

"Do you remember the proposal, Faith?"

His tone was calm and soothing.

The scent of cinnamon and man filled Faith's nose. "Yes, I–I remember," she stammered.

She could feel the warmth of Evan's hand as he took hers and held it gently. "May I have this dance?"

He was so close she thought she could taste the cinnamon in his mouth. She knew what he asked of her.

"Yes," she told him, finding strength in the one word. "Forever."

The embrace that followed told her how happy she had made Evan. He would never know the extent of her own happiness. She could only attempt to show him in the years that followed. Then she felt his lips on her cheek. She moved her own lips toward his until they touched, briefly, yet fully. The warmth of the kiss spread through her face, down her neck, throughout her being.

"I'll get my gear," Barnabas said, with a slight lilt in his voice.

Faith felt Evan turn in her arms, apparently toward the older man.

"Where are you headed?" he asked.

"I'm going to accompany this young lady on her trip to the fort," Barnabas said.

"What?" Charity exclaimed, clapping her hands so sharply that the horse reared back and whinnied.

Sits Alone reached for the reins and steadied the horse.

"But you're not taking these folks' good horse. We'll loan you one of our pack mules," Barnabas said.

"I'm not going anywhere!"

"Oh, yes, you are," he told her.

In the next moment Faith heard Charity and the horse being led away.

"Pa!" Charity shouted. "Tell this oaf to let me go!"

"Sorry, Charity, but it's time you got out on your own," he told his daughter. "You don't want to follow rules anymore, so you need to see what it's like in the real world."

Faith could hear the sadness in her father's voice, along with the sternness. Then she heard him step away from the group with the pack mule, probably to unload what they had brought back from hunting.

"Come with me," Faith heard Evan whisper in her ear. "We have much time to make up." She allowed herself to be led back into the home.

twelve

Faith awoke the next morning to the warmth of a bright sun and a wonderful sense of a new beginning. She snuggled under the quilt, breathing deeply from its folds, and caught a faint whiff of her mother's familiar scent. How Faith wished her mother were there—to see her accomplishments and share her joy.

Just as she started to pray, Sits Alone entered the dwelling.

Faith sat up at once, embarrassed to be in bed when everyone else must be up and about already.

"I have gifts for you, my sister," the Indian woman told her. "I have been up many days, before the sun, to prepare these."

"Gifts? For me?" Faith leaned forward eagerly.

"I made them for you," she said, walking over to the bed.

The scent of sage, mingled with the strong musty odor of tanned hide, found its way to Faith's nose. Then she felt something heavy laid across her legs. The tips of her fingers began exploring, feathering first over the sections of hide, sewn together at various points with even, straight stitches. Next she moved over the delicate beadwork, again, she could tell, of fine craftsmanship.

"It is a dress of white buckskin, for your wedding ceremony. I smoked it in sage, your favorite scent."

"Oh!" Faith gasped. "It is absolutely beautiful!"

"The right sleeve is longer than the left. You can feel the strap over the left shoulder—"

Faith swung her legs over the edge of the bed and started to slip the dress over her petticoat. Sits Alone stepped over to help her. When she had it on, Faith found the strap over the left shoulder, and the hide doubled back, hanging down from the chest in a loose flap. At the edge of this wave of hide was a line of quills, running diagonally from the right shoulder, around the body, ending below the left shoulder blade. The shift ended just below her knees.

Suddenly, she felt pressure on her ankle. Sits Alone had lifted Faith's right foot and was placing on it a soft, pliable shoe—a moccasin. She fit a moccasin onto the other foot.

Faith padded over to the rocker, enjoying the softness of the clothing, so unlike what she was used to wearing. "These feel wonderful!"

"You must take them off now," Sits Alone told her.

"Oh, but I want to wear them."

"I have others for you to wear."

"You do?"

Faith made her way back to the bed and let her friend work the shift and moccasins off. Then, without assistance, Faith slipped into another set of buckskin clothing which felt equally comfortable, but without the quills and beadwork. Once again she crossed to the rocking chair and sat down. Suddenly the expression on her face changed.

"Are you not pleased?" Sits Alone asked.

"Oh, yes!" Faith pressed her hand against her right

cheek, feeling its heat through her fingers.

"Then what makes you sad?"

Faith hesitated. "I–I was just wondering when a wedding will take place." Her voice was nearly a whisper. "Where is a church? Or a preacher?"

"This is not a thing for you to bother over," said her friend. "To everything there is a season."

"You learn well," she said, smiling.

Sits Alone had remembered the passage she had taught her in Ecclesiastes 3:1–8. *And a time to every purpose under the heaven,* Faith added in her thoughts.

"Do you want to know what I know now?" she asked, not waiting for an answer. "I understand what the Father has been trying to tell me. Do you remember Isaiah 42:16?"

"Yes," came the soft reply.

Faith repeated the passage. " 'And I will bring the blind by a way that they knew not; I will lead them in paths that they have not known: I will make darkness light before them, and crooked things straight. These things will I do unto them, and not forsake them.' "

"Yes," Sits Alone said again.

"When my family was killed, the Father was with me. I had to travel on paths that I have not traveled on before. And, as you taught me, that was another path: my independence. My darkness is not the same darkness as is in the souls of some men. I am blind, but I have the Light inside me. Understanding is Light. Ignorance is my darkness. And the Light is my Lord. With the Light as my guide, I can *see* the path I must travel on. So I will choose the straight over the crooked. The Light is Truth, and Truth will never forsake me."

A deep silence fell over the room. Faith moved in her seat. The feel of cool, soft buckskin against her skin provided unexpected comfort. Then she heard quiet crying.

"What is it?" she asked, breaking the silence.

"I have learned something as well," the older woman said. "That verse is meant for me too, for I have not understood my own suffering for a long time." She hesitated. "And I am so thankful to have a sister now. You I call 'One With the Wind' are a gift to me. From Father."

"And you are a gift to me," Faith assured her.

"Let us begin the day," Sits Alone said. "We have much to do."

Rising from the rocker, Faith crossed to the hearth where the water bucket rested. She gripped the pail's handle and made her way outside and down to the river. Next she would help Sits Alone crush dried venison with the *mano* and *metate* and make it into the succulent *pemmican* they had all grown to enjoy. They had already gathered berries and had the tallow ready to blend into the mixture. After that, they would have to peel and slice prairie turnips to add to the rabbit stew for the evening meal.

Faith could hear the voices of Doris and Virgil as they too worked the day away. Virgil was preparing logs for another home, their home, to be built soon. Doris was singing as she scrubbed laundry over a washboard that rested inside a large tub, scouring the clothes with soap and then rinsing them in the river. Faith listened for a moment as they called back and forth to each other in a playful manner while they worked. Then she went inside.

Faith was making the beds when Evan entered. She felt his arms wrap around her. All her dizzying thoughts disappeared inside his embrace.

"I love you, Faith," he whispered.

Fresh cinnamon entered her nose, and she drew in a deep breath, loving the scent and the man from which it came.

"I love you too, Evan. So very much."

"Do you now?" he asked, with a chuckle.

"Of course! I have always loved you, Evan. You know that!"

He laughed harder. "I'm teasing you, little one. Don't you know that?"

She felt a tightness in her stomach. "Oh. I'm sorry. I wasn't aware—"

"Stop now," he said, his voice low, calm. He drew her closer. "Of course I know you love me. But do you know how much I love you?"

"I guess so," Faith said hesitantly, not sure how to respond. Since she couldn't see his features, she didn't always understand his meaning when he was teasing her. Soon, though, she knew she would come to know Evan so well that she would catch things in his voice and even in his silence.

"Faith, my beloved, I love you so much it hurts. Do you understand that?" He didn't wait for her to answer. "I always have and always will. You were always meant for me, and the Lord knew that. I needed patience, though."

"Yes, but my sister didn't help matters any," Faith told him. "And I'm sorry for that."

"No need for you to be sorry." He reached out and touched her face gently. "It wasn't your fault."

"Everything worked out for the best." Faith leaned into his touch, loving this man more with every breath she took. She touched his face with her fingertips, running them over his eyes, nose, lips, and then the deep cleft in his chin.

She heard Evan shuffle his feet; then he pulled away.

"What? What is it?"

"Can we sit down and be together for awhile? I need some time with you."

Faith walked over to the rocking chair and sat down. She heard him drop to the buffalo rug in front of her. They talked well into the early morning hours, and no one interrupted them.

&

Three days came and went. Everyone was eating the afternoon meal when Barnabas returned. Charity wasn't with him.

"I smell someone else," Faith whispered in Evan's ear. She had just swallowed a bite of flat corn bread Sits Alone had fried for them. She breathed the strong scent of horse flesh and sweat.

Nobody said a word as Barnabas dismounted and joined the group inside.

"Who's with him?" Faith asked Evan.

Evan said nothing but waited instead for Barnabas to speak.

"Today's a special day," he told them. "It's special because we're going to make it special."

"But where is my sister?" Faith asked. "Is she okay?"

"She's fine. The army gave her a job as laundress—not head laundress. She'll have to work her way up to that. But the girl is fine."

Faith relaxed. "What do you mean about today being a special day?" *What is going on?* she wondered.

"I want to introduce Chaplain Mack Turnbow. He arrived at the fort the same time we did. He's agreed to marry the four of you." He paused, slapping his thigh. "What do you say to that?"

Faith didn't know what to say. She hadn't been expecting this. Then she heard Doris and Virgil whispering across the table.

"To everything there is a purpose," Faith said out loud, remembering her earlier conversation with Sits Alone.

Evan leaned close to her ear and whispered, "Yes."

"Yes," Faith said out loud.

"Us too," said Virgil.

Suddenly Faith felt pressure on her arm.

"Let us go. We must get ready!" Sits Alone sounded out of breath.

Faith let the Indian woman lead her to the inner room, with great excitement rising in her heart. She changed into the wedding garments Sits Alone had so lovingly created for her, since her mother's wedding dress had been destroyed in the fire. She heard Doris and Virgil in the loft, apparently preparing for the ceremony that awaited the four of them.

Within the hour they had gathered on the banks of the Rio Grande River, near the trees and flowers that grew in abundance. Faith and Evan, with Doris and Virgil, stood at the river's edge, with the preacher in front. Evan quietly slipped something behind Faith's ear, which she later learned was a beautiful yellow columbine.

Then the preacher began. "Dearly beloved, we are assembled here in the presence of God, to join this man and this woman—and this man and this woman—in holy marriage, which is instituted of God, regulated by His commandments, blessed by our Lord Jesus Christ, and to be held in honor among all men."

Faith listened as he spoke, his voice strong and commanding. She heard his words, "to cherish a mutual esteem and love; to bear with each other's infirmities and weaknesses; to comfort each other in sickness, trouble, and sorrow; in honesty and industry to provide for each other, and for their household, in temporal things; to pray for and encourage each other in the things which pertain to God; and to live together as the heirs of the grace of life."

He asked if any man was against the union of these couples. Then he asked God to "present these servants that they may be truly joined in the honorable estate of marriage, in the covenant of their God, and that they may be enriched by grace, to enjoy the comforts, undergo the cares, endure the trials, and perform the duties of life together as becomes Christians, under heavenly guidance and protection, through our Lord Jesus Christ." And then he said, "Amen."

He asked the women if they would accept the men in the holy bond of marriage.

Both said, "Yes," simultaneously.

Then he asked who would give these women away.

Mr. Lambert answered, "I will," for both.

Faith heard Doris crying and reached for her hand, holding it tightly throughout the rest of the ceremony.

The soon-to-be husbands were then asked if they would accept the women in the holy bond of marriage. They each responded favorably.

"As long as you all shall live," the preacher said.

He asked each of them to repeat the phrase that ended, "In sickness and in health, as long as we both shall live."

Then Sits Alone walked up to the couples and placed rings made of sinew and beads in their hands. After the rings were exchanged, "in token and pledge, and of constant faith and abiding love," the preacher said a few more words.

He then placed his hands over the clasped hands of all four of them, held in front, and stated, "Whom therefore God hath joined together, let no man put asunder."

As the couples embraced and kissed, the preacher gave a benediction. "The Lord bless you and keep you: the Lord make his face to shine upon you and be gracious unto you: The Lord lift up his countenance upon you, and give you peace: All now, and in life everlasting. Amen."

For the first time in her life Faith felt complete, something she had never thought she would feel.

She felt Evan's breath against her neck.

"Come with me, my love," he told her. "I have prepared a place for us, where we can meet under the stars and the moon, beneath the heavens—where we can join together as one. We are one now."

"Yes, my love. And I will be with you for eternity."

Her heart skipped a beat, and her stomach fluttered. Now she would know what it felt like to be a woman.

The Lord had not forsaken her. He would be with her always. She knew that. In her heart she thanked Him, as she walked to a distant place Evan had earlier prepared for them, apparently while she was getting ready for the wedding.

"I love you, Evan."

"I love you, my beloved."

And they became as one.

A Letter To Our Readers

Dear Reader:

In order that we might better contribute to your reading enjoyment, we would appreciate your taking a few minutes to respond to the following questions. We welcome your comments and read each form and letter we receive. When completed, please return to the following:

Rebecca Germany, Fiction Editor
Heartsong Presents
PO Box 719
Uhrichsville, Ohio 44683

1. Did you enjoy reading *One with the Wind* by Kelly R. Stevens?
   - ❑ Very much! I would like to see more books by this author!
   - ❑ Moderately. I would have enjoyed it more if

   _____

   _____

2. Are you a member of **Heartsong Presents**? Yes ❑ No ❑
   If no, where did you purchase this book?_____

   _____

3. How would you rate, on a scale from 1 (poor) to 5 (superior), the cover design?_____

4. On a scale from 1 (poor) to 10 (superior), please rate the following elements.

   _____ Heroine   _____ Plot

   _____ Hero   _____ Inspirational theme

   _____ Setting   _____ Secondary characters

5. These characters were special because _____

_____

_____

6. How has this book inspired your life? _____

_____

_____

7. What settings would you like to see covered in future
   Heartsong Presents books? _____

_____

_____

8. What are some inspirational themes you would like to see
   treated in future books? _____

_____

_____

9. Would you be interested in reading other **Heartsong
   Presents** titles?   Yes ❏   No ❏

10. Please check your age range:
    ❏ Under 18   ❏ 18-24   ❏ 25-34
    ❏ 35-45   ❏ 46-55   ❏ Over 55

Name _____

Occupation _____

Address _____

City _____ State _____ Zip _____

Email _____

the Sewing Circle

Edna Tidewell is a Titus woman. Following the Biblical injunction to mentor younger women in Christ, this pastor's wife sets aside each Tuesday afternoon for the young women of Hickory Corners, Ohio. During "Tea with Mrs. T," the ladies do personal sewing or hold a quilting bee while Mrs. Tidewell reads aloud from the Scriptures. Discussion, tea, and sweet conversation round out each afternoon.

These special get-togethers could have far-reaching effects on four young ladies.

Relax along the banks of the historic Ohio River and watch four searching young women transform into stable servants of the Lord. Their stories will warm your heart and inspire your soul.

paperback, 352 pages, 5 ³⁄₁₆" x 8"

♥ • ♥ • ♥ • ♥ • ♥ • ❤ • ♥ • ♥ • ♥ • ♥ • ♥

♥ • ♥ • ♥ • ♥ • ♥ • ❤ • ♥ • ♥ • ♥ • ♥ • ♥

·······Presents·······

Great Inspirational Romance at a Great Price!

Heartsong Presents books are inspirational romances in contemporary and historical settings, designed to give you an enjoyable, spirit-lifting reading experience. You can choose wonderfully written titles from some of today's best authors like Peggy Darty, Sally Laity, Tracie Peterson, Colleen L. Reece, Lauraine Snelling, and many others.

When ordering quantities less than twelve, above titles are $2.95 each.
Not all titles may be available at time of order.